84th Conference of the New Zealand Society of Animal Production (NZSAP 2024)

Proceedings of the New Zealand Society of Animal Production
Volume 84

Omaru, New Zealand
5-7 November 2024

ISBN: 979-8-3313-2213-7

Printed from e-media with permission by:

Curran Associates, Inc.
57 Morehouse Lane
Red Hook, NY 12571

Some format issues inherent in the e-media version may also appear in this print version.

New Zealand Society of Animal Production (Inc.) (NZSAP)
c/o Kate Crookston
PO Box 955
Cambridge, New Zealand

www.nzsap.org.nz

Additional copies of this publication are available from:

Curran Associates, Inc.
57 Morehouse Lane
Red Hook, NY 12571 USA
Phone: 845-758-0400
Fax: 845-758-2633
Email: curran@proceedings.com
Web: www.proceedings.com

TABLE OF CONTENTS

New Zealand Journal of Animal Science and Production 2024 Vol 84. 1-3

BRIEF COMMUNICATION: Liveweight, body composition, carcass traits and meat quality of artificially reared milk-fed Friesian-Jersey crossbred heifer and bull calves

SA McCoard[1*], E Pavan[1,2,3], K Taukiri[2], R Zhang[2] & CE Realini[2]

[1]AgResearch Limited, Grasslands Research Centre, Private Bag 11008, Palmerston North 4442, New Zealand;
[2]Food and Technology & Processing Team, Te Ohu Rangahau Kai, AgResearch Limited, Palmerston North 4474, New Zealand;
[3] Unidad Integrada Balcarce (Estación Experimental Agropecuaria Balcarce, Instituto Nacional de Tecnología Agropecuaria—Facultad de Ciencias Agrarias, Universidad Nacional de Mar del Plata), CC 276, Balcarce 7620, Argentina.
**Corresponding author email: sue.mccoard@agresearch.co.nz*

Abstract

Novel approaches are required by the dairy industry to identify productive endpoints for surplus calves beyond 4-day-old bobby calves. The objective of this study was to characterise carcass weight, body composition, and meat quality attributes of ~11-week-old milk-fed dairy bull and heifer calves. Mixed-sex Friesian-Jersey crossbred dairy calves (n=73) were artificially reared. Calves were weighed at birth and slaughter. Carcass weight, heart girth, and the weight of selected organs and muscles were recorded. Meat quality parameters were evaluated including muscle pH, colour, and intramuscular fat percentage (IMF). Bulls had greater average daily gain (P=0.003), slaughter weight (P=0.007), and carcass weight (P=0.022) than heifers. IMF % (P=0.038), redness (P=0.009), yellowness (P=0.041), and chroma (P=0.011) were greater in heifers than bulls, while pH, lightness, and hue angle were similar (P>0.05). This study provides new information to inform the potential utilisation of milk-fed calves for novel meat products.

Keywords: Cattle; growth, meat quality, milk-fed veal

Introduction

The Dairy Tomorrow Strategy (DairyNZ et al. 2017) outlined a vision and strategy for the future of the New Zealand dairy sector involving a commitment to a high level of animal care that provides all animals a life worth living. Practices such as processing very young animals (e.g., 4-day-old bobby calves) are being challenged by increasing consumers' interest in the ethical standards of food systems, and concerns around the processing of young animals are expected to continue to increase (Bolton & Von Keyserlingk 2021). However, rearing all the surplus calves is a major challenge for the industry as there is not sufficient land available without displacing other land uses.

At least 74% of New Zealand beef produced originates in the dairy industry (Edwards et al. 2021). On average, it is estimated at the herd level that 28% of calves are retained as replacements, 12% are reared on-farm for beef, 15% are sold for beef rearing, 35% are sold to process as "4-day-old bobby calves", 5% are euthanised and 4% are born dead or died (Edwards et al. 2021). Identification of new opportunities to utilise the large number of non-replacement calves from the New Zealand dairy cow industry and eliminate the processing of bobby calves is important to maintain the social licence to operate.

Current initiatives to explore alternative meat value chains for these calves have focused on rosé veal from 8- to 12-month-old animals. Our prior research in dairy sheep lambs indicates that a 3-week-old milk-fed meat product has similar meat quality characteristics and intramuscular fat content to ~3-4 month-old grass-fed lambs and is highly desirable by consumers (Pavan et al. 2022). Similar data for milk-fed veal calves from New Zealand pasture-based rearing systems, to support exploration of potential novel meat product alternatives for this class of animal are not available.

The objective of this study was to generate baseline information on carcass weight, body composition (muscle, fat, organs), carcass yield and meat quality characteristics in 10-12 week-old milk-fed dairy calves reared in a pasture-based management system.

Materials and methods

All animal manipulations in this study were approved by the AgResearch Animal Ethics Committee (Approval number AEC808). The animals used in this study were sourced from another experiment evaluating a nutritional supplement ("treatment") where there was no effect of treatment on calf parameters. Seventy-three mixed-sex (35 heifers and 38 bulls) crossbred calves were born in Spring 2023 to mixed-age Kiwicross dams (Friesian sire) calved indoors on straw. The calves were removed from the dams within 10 hours of birth and allocated to one of 9 indoor mixed sex rearing pens (3 pens/treatment) bedded with straw on the farm of origin (Dairy 4, Massey University, Tennent Drive, Palmerston North, NZ). All calves were offered three 2L feeds of warm pooled gold colostrum (Brix >24) from their dams over the first 36 hours post-collection. Calves were then offered warm transition milk (2 × 2L per calf per day) for the next 9 feeds and then transitioned over 2 days onto a warm calf milk replacer (CMR, AnCalf, NZAgBiz, NZ) mixed at 150 g/L according to the manufacuter's recommendations. At approximately 10 days of age, all calves were transported for ~30 minutes to the AgResearch Aorangi Research Farm (Lockwood Road, Palmerston North, NZ). Calves were randomly allocated to pens (~12 calves/pen) and fed 2 × 2L of warm CMR per day (600 g DM/d mixed at 150 g/L) for ~10 days in a north-facing 3-sided pole shed bedded with wood chips. Calves were then transitioned outdoors over 5 days by offering free indoor-outdoor access to ~~a~~ pasture and offered 2 × 2.5L CMR per day (788 g DM/d mixed at 175 g/L) for 5 days (~25/group – "cohort"). At ~25 days of age, calves were then moved in their groups to one of three sheltered pasture paddocks and offered 2 × 2L of warm CMR per day (~800g DM/d; mixed at 200 g/L) for 9 days. Calves were transitioned to once-a-day feeding (final volume of 4.5L/hd/d) over 5 days (1049 g DM/d; mixed at 233 g/L). Concentration of CMR was increased over time to achieve target dry matter intake with a lower milk volume to stimulate solid feed intake. Calves were maintained at this feeding level until the end of the study. All calves were fed using compartmentalised feeders (Stallion Plastics, Palmerston North, NZ) and individually monitored to ensure each calf received its allocated feed. Calves were offered *ad libitum* grain-based calf starter (20% protein pellets, Seales Winslow, NZ) from arrival at Aorangi Farm.

All calves were weighed at birth and at the end of the study to calculate average daily gain (ADG). All calves received a trace element injection (Multimin®, Virbac, Hamilton, NZ) on arrival to Aorangi Farm (~10 days of age) and a clostridial vaccine (Ultravac 7:1, Zoetis, Auckland, NZ) at ~5.5 weeks of age and a booster 4 weeks later. No calves were disbudded or castrated.

All calves were fed at noon on the day prior to humane killing. Calves in each "cohort" were humanely killed by captive bolt stunning, pithing, and exsanguination at a similar age. Weight of the femur, tongue, selected organs (heart, liver, kidney, spleen, thymus) and perirenal fat were recorded. The carcasses were hung and chilled overnight at 4°C. On the following day, the cold carcass weight was recorded. Temperature and pH were measured at 24 hours. The *M. Longissimus thoracis et lumborum* (LTL) and *M. Semitendinosus* were removed from the right side of the carcass, weighed, and subsamples collected for meat quality assessments as previously described (Pavan et al. 2022). Instrumental colour (Minolta Chroma Meter) was measured after 30 min bloom time on the exposed LTL surface. The intramuscular fat (IMF) percentage of loin samples (LTL) was measured based on total fatty acids detected in the muscles using a direct trans-methylation method (Agnew et al. 2019).

Data was analysed using the SAS System and a mixed model including sex, treatment, and their interaction as fixed effects and pen nested within treatment, animal nested within pen and treatment, and breed of dam as random effects. Least squares means were compared using Tukey's test. Data are presented as means and standard errors of the mean.

Results and discussion

No calves died during this study and one calf was treated for an infection at the ear tag site. No other animal health issues were evident. Bull calves had greater ADG, slaughter weight, carcass weight and yield and lower perirenal fat mass compared to heifers, while no other differences in body composition were observed (Table 1). The higher growth rate of bull compared to heifer calves is consistent with previous observations in Angus and crossbred calves pre-weaning (Woods et al., 1990) but contrasts those in Friesian calves (McCoard et al. 2019). Reasons for the differences between studies are unclear but may be related to the feeding system used or breed-specific effects.

Meat from heifers showed higher colour values for redness, yellowness, and chroma indicating a more vibrant meat colour compared with bulls. This may be due to differences in muscle fibre characteristics and fat content between heifers and bulls (Kučević et al., 2019). In this study, heifers had greater IMF percentages (0.78 vs 0.69%, P=0.038) compared to bulls, which is consistent with observations in older animals (Hughes et al. 2019; Kučević et al., 2019; Tan and Jiang 2024). There were no differences in muscle pH or colour lightness and hue angle between heifers and bulls.

This study provides baseline information on growth, carcass and meat quality attributes of heifer and bull milk-fed dairy animals, to support future studies to determine the potential of this class of animals for novel milk-fed veal products from pasture-based management systems with low chemical inputs and minimal painful procedures. Future work on consumer/society, welfare guidelines for their production and opportunities to improve eating quality through nutrition would provide further insights into the potential opportunities for a novel meat product from this class of animal to contribute to new opportunities for the large number of non-replacement calves from the dairy cow industry.

Table 1. The effect of calf sex on live weight, average daily gain, body composition and meat quality attributes of FxJ dairy calves slaughtered at ~11 weeks of age.

Parameter	**Heifer (n=35)**	**Bull (n=38)**	**P-value**
Average daily gain (g/d)	798 ± 14	850 ± 13	**0.003**
Slaughter weight (kg)	101.9 ± 1.5	105.5 ± 1.4	**0.045**
Slaughter age (d)	80.3 ± 0.9	78.3 ± 0.9	**0.003**
Body composition:			
Carcass weight (kg)	48.2 ± 1.0	51.1 ± 0.9	**0.013**
Carcass yield	47.3 ± 0.4	48.5 ± 0.4	**0.048**
Heart girth at slaughter (cm)	107.8 ± 0.6	108.7 ± 0.6	0.234
Femur weight (g)	1.13 ± 0.02	1.06 ± 0.02	0.063
Heart weight (g)	627 ± 13	643 ± 12	0.379
Liver weight (kg)	2.20 ± 0.05	2.30 ± 0.05	0.121
Kidney (x2) weights (g)	482 ± 13	474 ± 12	0.585
Perirenal fat (g)	463 ± 31	396 ± 29	**0.038**
Spleen (g)	544 ± 69	564 ± 66	0.063
Thymus (g)	432 ± 14	454 ± 13	0.245
Tongue (g)	303 ± 8	308 ± 7	0.467
M. Longissimus thoracis lumborum weight (g)	1150 ± 28	1179 ± 26	0.393
M. Semitendinosus weight (g)	431 ± 12	440 ± 12	0.427
Meat quality attributes (*M. Longissimus thoracis*):			
pH at 24 hrs postmortem	5.7±0.03	5.7±0.03	0.591
L* (lightness)	44.2±0.53	44.1±0.52	0.839
a* (redness)	12.3±0.24	11.5±0.23	**0.009**
b* (yellowness)	8.7±0.20	8.1±0.19	**0.041**
Chroma	15.1±0.28	14.1±0.27	**0.011**
Hue angle	34.8±0.62	35.4±0.59	0.419
Intramuscular fat %	0.78±0.031	0.69±0.030	**0.038**

Acknowledgements

The authors sincerely thank the Massey University Dairy 4, AgResearch Aorangi Research Farm and Ulyatt Reid Large Animal Facility staff for the management and care of the animals during the study and technical support. This study was funded by the AgResearch Strategic Science Investment Fund (A25765 and A27235) and Ballance AgriNutrients and the Ministry of Primary Industries "Future Ready Farms – Together, creating a lower emissions Future" Sustainable Farming Futures Fund program.

References

Agnew MP, Craigie CR, Weralupitiya G, Reis MM, Johnson PL, Reis MG 2019. Comprehensive evaluation of parameters affecting one-step method for quantitative analysis of fatty acids in meat. Metabolites 9(9): 189.

Bolton SE and von Keyserlingk AG 2021. The dispensable surplus dairy calf: Is this issue a "wicked problem" and where do we go from here? Frontiers in Veterinary Science 8: 660934.

DairyNZ, Federated Farmers, DCANZ, Dairy Women's Network. 2017. Dairy Tomorrow: The future of New Zealand dairying. https://www.dairytomorrow.co.nz/

Edwards JP, Cuthbert S, Pinxterhuis JB, McDermott A 2021. The fate of calves born on New Zealand dairy farms and dairy farmer attitudes towards producing dairy-beef calves. The New Zealand Journal of Animal Science and Production 81: 179-185.

Hughes J, Clarke F, Li Y, Purslow P and Warner R 2019. Differences in light scattering between pale and dark beef longissimus thoracis muscles are primarily caused by differences in the myofilament lattice, myofibril and muscle fibre transverse spacings. Meat Science 149: 96-106.

Kučević D, Papović T, Tomović V, Plavšić M, Jajić I, Krstović S, Stanojević D 2019. Influence of farm management for calves on growth performance and meat quality traits during fattening of simmental bulls and heifers. Animals 9(11): 941.

McCoard S, Heiser A, Lowe K, Molenaar A, Maclean P, Johnstone P, Leath P, Hoskin SO, Khan MA 2019. Effect of weaning age on growth, mammary gland development, and immune function in Holstein Friesian calves fed conserved alfalfa (FiberStart). Journal of Dairy Science 102: 6076-6087.

Pavan E, McCoard S, Agnew M, Zhang R, Taukiri K, Farouk MM, Realini CR 2022. Effects of dairy lambs' rearing system and slaughter age on consumer liking of lamb meat and its association with lipid content and composition. Foods 11, 2350.

Tan Z and Jiang H 2024. Molecular and cellular mechanisms of intramuscular fat development and growth in cattle. International Journal of Molecular Sciences 25, 2520. 2024.

Woods BL, Bradley NW, Schillo KK, Lowry SR 1990. Effects of nutrition, sex of calf and breed type on response to zeranol: preweaning growth. Journal of Animal Science 68:919-922.

New Zealand Journal of Animal Science and Production 2024 Vol 84. 4-6

BRIEF COMMUNICATION: Rate of Genetic Gain for Methane Emissions in a Maternal Production Flock

F Booker*, McEwan, JC & SJ Rowe

AgResearch Ltd, Invermay Agricultural Centre, 176 Puddle Alley, Mosgiel,
*Corresponding author email: fern.booker@agresearch.co.nz

Abstract

Global warming attributed to methane emissions from livestock presents a significant global challenge to the agricultural sector. Although opportunities exist in intensive systems for feed additives, the only proven tool for permanent mitigation at a national level, across sectors, is animal breeding. It has been reported that enteric methane emission is a heritable trait, and that breeding is an effective mitigation strategy. The impacts of incorporating methane into existing breeding schemes are less understood. Selection for reduced methane emissions (measured by portable accumulation chamber; PAC) along with increased productivity (NZMW+FEC+DAGS+MEAT index) was implemented in 2017 in the Woodlands 2638 flock. The economic weighting for methane was NZ$100/tonne GWP100 CO2e. The economic weights for existing traits were unchanged. Breeding value trends per year for flock 2638 and the Dual Purpose industry average for born 2018 to 2023 progeny were estimated as -0.95 vs -0.01%, -0.2 vs 0.06%, 0.18 vs 0.12kg, 133 vs 89 cents, and 185 vs 110 cents for PACCH4gBV%, PACCO2gBV%, LW8gBV, NZMW and NZMW+FEC+DAGS+MEAT indexes respectively. Gene flow calculations suggest that methane emissions could be reduced by 27% by 2050 while increasing per head productivity by $51.80. Continuation with current industry progress would achieve a 2.8% reduction in methane emissions and $30.80 increase in per head productivity. In summary, methane emissions can be reduced while also increasing productivity faster than current industry progress.

Keywords: methane; breeding; sheep; PAC

Introduction

Methane (CH4) emisions from ruminants is one of the largest sources of anthropogenic CH4 globally (Schaefer et al. 2016). The New Zealand government has set a target of a 10% reduction in methane emissions by 2030, with further reduction of 24-47% by 2050 (Paris Agreement 2015). There are opportunities in more intensive farming systems to reduce methane emissions through the use of feed additives, but in New Zealand's extensive pastoral-based system these additives are less useful. Methane emissions from sheep determined using Portable Accumulation Chambers (PAC) have been shown to be moderately heritable (Jonker et al. 2018), giving a permanent and effective mitigation strategy.

Research suggests that breeding for lowered methane per kg DMI does not have negative effects on other production traits, and may in fact have a positive impact on some other traits such as fleece weight (Hickey et al. 2022). To date, there is very little known about the effects of incorporating methane selection into breeding schemes whilst still selecting for genetic gain in other production traits.

Materials and methods

Sheep Improvement Ltd (SIL) Flock 2638 is a maternal genetics research flock based at AgResearch Woodlands, Southland, consisting of 750 ewes plus replacements. It uses 20 rams per year, of which 2 are repeat matings and 4 are link rams from other flocks. The flock began primarily as a Coopworth flock, but over recent years has developed into a maternal composite. Production traits on this flock have been collected and recorded as per the SIL Best Practice Guide (Beef and Lamb Genetics 2016). Methane emissions were measured through PAC as per the protocol described by Jonker et al. (2018). Flock 2638 has been selected on a custom economic index incorporating methane, commencing with born 2017 lambs. The index is NZMW+FEC+DAGS+MEAT+PAC. (Anon 2022). To incorporate PAC breeding values into an economic index, methane has been given a cost of $100 per tonne of GWP100 CO2 equivalents. This results in a cost of $6.82 per gram/day of PAC methane breeding value (PAC CH4 gBV g/d) or $0.51 per percent reduction when expressed as PAC CH4 gBV%. Average breeding values and indexes for the born 2017 to 2023 progeny for flock 2638 and all dual purpose flocks were retreived from SIL in June 2024 (Job No 41674) and the genetic trends and averages estimated by linear regression.

Results and discussion

Selecting for reduced methane

Figure 1 shows the means and change that has been made in the PAC methane emission gBV% in flock 2638 relative to all dual purpose recorded sheep since 2017. Starting from similar values, flock 2638 has been declining by -0.95% per year on a per animal basis while for dual purpose animals the decline has been only -0.01% per year. For the production component of the selection index (NZMW+FEC+DAGS+MEAT) flock 2638 born 2017 progeny averaged 2937 cents in 2017 compared to 1347 cents for the average dual purpose animal. Over the next 6 years the trend was an increase of 185 cents per year for flock 2638 compared to 110 cents per year for the average dual purpose animal. A difference in the rate of productivity gain of 67%.

Many New Zealand dual purpose sheep are selected only on a simplified production index, New Zealand maternal worth (NZMW) incorporating reproduction, survival, growth and adult size traits. When comparing these traits, flock 2638 born 2017 progeny averaged 2294 cents in 2017 compared to 1256 cents for the average dual purpose animal. Over the next 6 years the trend was an increase of 133 cents per year for flock 2638 compared to 89 cents per year for the average dual purpose animal. A difference in the rate of gain in favor of flock 2638 of 49%. The current PACCH4gBV (grams of methane emitted per day per animal from PAC data)is expressed on a per animal basis rather than a per kg of dry matter intake (DMI) basis so changes in CO_2 emission breeding values (a reflection of intake) and liveweight are also of interest. A common industry statement is that selection for lower methane emissions will simply result in smaller animals.When comparing carbon dioxide emission breeding values, flock 2638 born 2017 progeny averaged -1.4% in 2017 compared to -0.1% for the average dual purpose animal. Over the next 6 years the trend was a decrease of -0.2% per year for flock 2638 compared to an increase of 0.06 per year for the average dual purpose animal. Although part of this difference could be changes in digestability, it suggests feed efficiency is unchanged or perhaps slowly improving. In contrast, when growth in the form of breeding values for liveweight at 8 months of age (LW8gBV) are examined, flock 2638 born 2017 progeny averaged 4.08kg in 2017 compared to 3.86kg for dual purpose animals. Over the next 6 years the trend was 0.12 kg per year for flock 2638 compared to an increase of 0.18kg per year for the average dual purpose animal. This suggests a modest decrease in the rate of gain in liveweight by the inclusion of not only methane but also the disease traits FEC and DAGS and MEAT in the flock 2638 selection index. However it did not result in a decrease in liveweight.

In summary the results indicate that it is possible to select for a reduction in methane emissions, while still maintaining genetic progress in other production and health traits at a rate greater than the industry is currently acheiving. The reasons behind this are not simply measuring methane via PAC, but also include the use of genomic selection, better breeding scheme design and greater emphasis on the selection index when selecting flock replacements.

Selective breeding to hit emissions reduction targets.
We next investigated if the rate of reduction in methane emmissions that has been seen in flock 2638 over the past 6 years was sufficient to acheive New Zealand's targeted reductions by the year 2050. Assuming the rate of reduction of –0.95% /year as has been shown in the flock so far will be maintained, and accounting for the genetic lag for use of rams in commercial flocks, this would result in a 27% decrease in commercial flocks methane emissions by the year 2050 while increasing per head productivity by $51.80. Continuation with current industry progress would achieve a 2.8% reduction in methane emissions and $30.80 increase in per head productivity. In summary, methane emissions can be reduced while also increasing productivity faster than current industry progress. While this analysis does not account for all factors it indicates that genetic selection, if appropriately applied, could contribute a substantial proportion of the currently proposed reduction in methane emissions from the sheep industry while also improving their productivity at a rate faster than currently being acheived and without any change in ewe numbers. The fact that these results have been obtained from a SIL recorded flock and using the current industry breeding evaluation system adds strength to the fact that this is possible using existing industry tools available to all New Zealand breeders. The challenge is its rapid adoption by the industry.

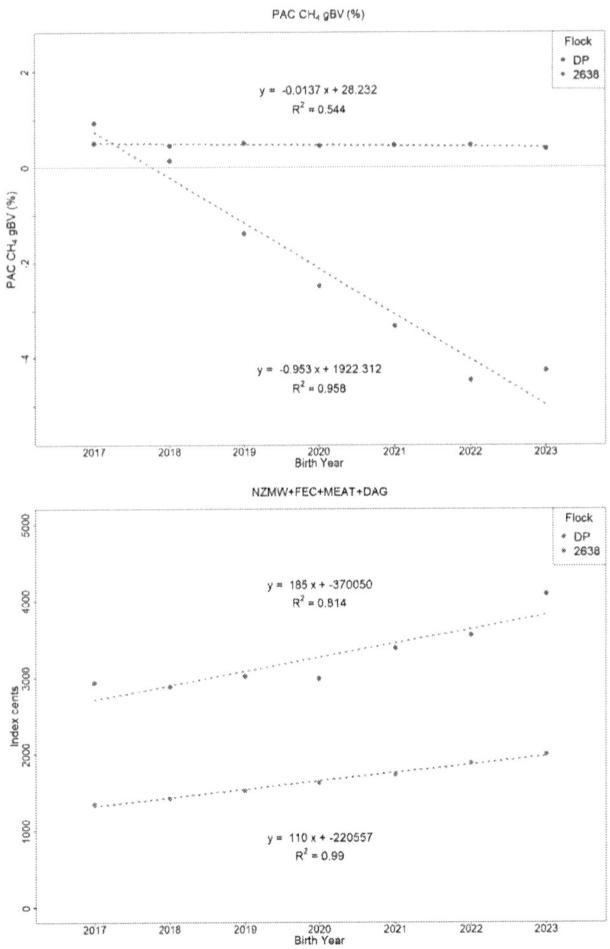

Figure 1. Plots of percentage change in average PAC CH4 breeding values (PACCH4gBV%) and NZMW+FEC+DAGS+MEAT index values (cents, NZMW+F+D+M) by birth year for flock 2638 and the New Zealand dual purpose flock average (DP).

Acknowledgements
Thanks to the Woodlands and Invermay staff who manage and record flock 2638, to the breeders who are now recording methane traits, and to B+LNZ Genetics for the breeding value estimates. Thanks to Tom Kelly for your support with generating figures.

References

Anon. Sheep industry Breeding objectives 2022 Available at https://www.sil.co.nz/files/1661310874929.pdf [accessed 21 June 2024]

Beef and Lamb Genetics 2016 Best Practice in Performance Recording for SIL Genetic Evaluations, Beef and Lamb Genetics https://www.sil.co.nz/technical [accessed 8 October 2024]

Hickey SM, Bain WE, Bilton TP, Greer GJ, Elmes S, Bryson B, Pinares-Patiño CS, Wing J, Jonker A, Young EA, Knowler K. 2022. Impact of breeding for reduced methane emissions in New Zealand sheep on maternal and health traits. Frontiers in Genetics 13: 910413.

Jonker A, Hickey SM, Rowe SJ, Janssen PH, Shackell GH, Elmes S, Bain WE, Wing J, Greer GJ, Bryson B, MacLean S. 2018. Genetic parameters of methane emissions determined using portable accumulation chambers in lambs and ewes grazing pasture and genetic correlations with emissions determined in respiration chambers. Journal of Animal Science 96(8): 3031-3042.

United Nations / Framework Convention on Climate Change. 2015 Adoption of the Paris Agreement, 21st Conference of the Parties, Paris: United Nations

Schaefer H, Fletcher SEM, Veidt C, Lassey KR, Brailsford GW, Bromley TM, Dlugokencky EJ, Michel SE, Miller JB, Levin I, Lowe DC. 2016. A 21st-century shift from fossil-fuel to biogenic methane emissions indicated by 13CH4. Science 352: 80-84.

New Zealand Journal of Animal Science and Production 2024 Vol 84. 7-10

BRIEF COMMUNICATION: Effects of measuring methane in pregnant ewes using portable accumulation chambers on pregnancy outcomes and lamb performance to weaning

J. Peers-Adams, T.P. Bilton, S.J. Rowe and J.L. Juengel

AgResearch Ltd., Invermay Research Centre, Puddle Alley, Private Bag 50034, Mosgiel 9053, NZ
*Corresponding author email: Jacqueline.Peers-Adams@Agresearch.co.nz

Abstract

The portable accumulation chamber (PAC) method provides an on-farm methane measurement over a short timeframe that is scalable to measure large numbers of animals. The aim of this study was to assess the safety of use of PAC in pregnant ewes. Methane production was measured in ewes (n=40 per group) using PAC at approximately 70 or 100 days of gestation. Control animals were either exposed to yarding and sorting only (Days 70 and 100) or exposed to yarding, sorting and restriction from feed as outlined in the standard operating procedure for PAC (Day 100 only). Ewe health was not differentially affected by PAC measurement. Confinement in PAC did not affect ($P>0.50$) lamb survival, with the proportion of lambs weaned relative to number at mid-pregnancy ranging between 0.817 to 0.898. Birth and weaning weights of the lamb were similar ($P>0.28$) between all groups (birth weight ranged between 4.8 to 5.1 kg, weaning weight ranged between 28.1 to 29.3 kg). Carbon dioxide levels in PAC were not elevated compared to what has previously been observed in non-pregnant ewes. Methane concentrations were calculated for the ewes around day 100 of pregnancy and the average methane production was 23.2 ± 5.8 g per day. In conclusion, PAC appears to be a safe method for measuring methane traits in ewes during the middle to later stages of pregnancy.

Keywords: PAC methane measurements; Pregnant ewes; Lambing performance

Introduction

Methane production by ruminant animals is recognised as a significant contributor to greenhouse gas emissions and global warming. Globally, enteric methane production from ruminant's accounts for around 6% of anthropogenic greenhouse gas emissions (Beauchemin et al. 2020). The contribution that enteric methane makes to New Zealand's greenhouse gas emissions is even higher, accounting for over 35 % of total emissions (Ministry for the Environment 2023). Reducing enteric methane emissions from ruminants is seen as essential to allow NZ, and the world, to meet emissions reduction targets.

A key aspect of developing approaches to reduce enteric methane production is being able to accurately measure methane production. This allows assessment of the effects of diet, feed additives, genetics, and other products on methane production to enable development of the most effective approaches for reducing methane. Multiple methods have been developed to measure methane production, with the portable accumulation chamber (PAC) being recognised for its ability to measure a large number of sheep in a short timeframe (Goopy et al. 2011; Jonker et al. 2018). These chambers can be scaled to include multiple units on a single trailer, making the measurement of 80+ sheep in a single day with a single trailer containing 12 units feasible. A single measurement of less than an hour in duration can be used to provide measurements of methane production suitable for use in genetic selection programmes, which are currently being commercially applied in New Zealand (Lawrence 2022).

Breeding females account for a significant proportion of farm greenhouse gas emissions. Additionally, ewes are pregnant for ~40% of the year. Thus, it is important to understand if pregnancy affects methane production and potential interactions between pregnancy and methane mitigation approaches. To do this, methods to measure methane production in pregnant ewes, without compromising the health of the dam or offspring, need to be identified.

Use of a PAC requires the animal to be held in a sealed environment for the duration of the time that the measurement is made (Goopy et al. 2011; Jonker et al. 2018). During this timeframe, oxygen is limited to that originally in the chamber at the time of sealing. Currently, PAC's have been primarily utilised on young growing lambs (prepubertal), with some adult male and non-pregnant females also being measured without incident. However, it is known that longer term oxygen deprivation, particularly later in gestation, can have negative impacts on the pregnant female, and particularly her offspring (Morrison 2008).

Our aim was to determine if a single PAC measurement would affect the ewe's or her offsprings' health. A secondary aim was to measure methane production in pregnant ewes and determine if number of foetuses affected methane production.

Materials and methods

The experiment was run in accordance with the New Zealand Animal Welfare regulations and approved by the AgResearch Animal Ethics Committee (application number 1912).

Ewes were selected from a group of mature (2.5-4.5 years of age) breeding ewes with recorded mating marks (recorded every Monday, Wednesday, Friday). A premating weight was obtained on the first day of the breeding period. Transabdominal ultrasound scanning was utilised to determine pregnancy status including litter size. All animals were managed as a single group on pasture and yarded at each PAC measurement for sorting of animals. Four groups of ewes (n=40 per group) were selected : ewes yarded but returned to pasture (65-77 days of gestation for first PAC measurement and 95-107 for second PAC measurement; Control), underwent a PAC measurement between 65-77 days of gestation (PAC mid-pregnancy), underwent a PAC measurement from 95-107 days of gestation (PAC later pregnancy), or were yarded, retained in the yards (thus restricted from access to feed) but not exposed to the PAC at 95-107 days of gestation (Feed restriction). All ewes underwent another transabdominal ultrasound scan approximately one month after the initial pregnancy scan. Following assigned treatment, all ewes were fed on pasture or crops to meet their requirements as assessed by experience farm staff. A fully recorded lambing was undertaken on these animals which included assigning lamb to dam and obtaining a birth weight and determining sex of the lamb within 24 hours of birth. Status of the lamb (alive or dead) was recorded within 24 hours of birth and at weaning. Weaning weight was also recorded.

Portable accumulation chambers were used to measure methane as previously described (Jonker et al., 2018). At each measurement timepoint, all ewes were measured for methane production on the same day. For each measurement day, a total of 6 rounds of measurements, with 6-7 animals per measurement round, were undertaken. Animal number per round were reduced from the typical 12 to allow more frequent monitoring of the pregnant animals given the uncertainty of their reaction to the chambers. Time off feed ranged from approximately 2 to 4 hours.

A total of 6 animals were removed from the analysis due to animal health problems associated with normal lambing (1 ewe from each group died before lambing) or lambing date was not recorded (1 ewe in the PAC mid-pregnancy group and 1 ewe in Feed restriction group). Foetal age for all groups was calculated using the observed mating mark. The following ewe traits were calculated. The proportion of lambs born relative to the number of lambs at scanning (PLB). The proportion of lambs born alive relative to the number of lambs at scanning (PLBA). The proportion of lambs weaned relative to the number of lambs at scanning (PLW). The proportion of lambs weaned relative to the number of lambs born alive (PLLW). The power to detect a 20% drop in survival for a treatment relative to the control was between 61.2% (PLW) and 82.9% (PLB).

Lambs were only fostered to other ewes if an experienced shepherd determined that leaving the lamb with the mother was likely to result in the lamb's death (dam abandoned lamb, dam was not providing sufficient milk for lamb). For the purpose of the maternal traits, lambs that were removed from the dam were considered to have died between birth and weaning. The following lamb traits were assessed. Survival of a lamb born alive to weaning (LSW), the weight of the lamb around birth whether the lamb was alive or dead (BWt) and the weight of the lamb at weaning (WWt). For lamb traits, a lamb that was fostered onto another ewe was retained in the data set. Difference in lambing date were calculated by determining the deviation (in days) in the birthdate relative to the mean birthdate of all animals (bdev).

All statistical analyses were carried out in R. Differences in age, number of foetuses present at mid-pregnancy scanning, liveweight of the ewe just prior to joining with the ram, foetal age, and the birth date deviation from average birth data between groups was analysed using ANOVA (Table 1). A generalised linear model (binomial) was fitted for the ewe traits, with a generalised linear mixed model (binomial) used for the lamb trait LSW and linear mixed model (gaussian) used for the BWt and WWt lamb traits. For the maternal traits PLBA, PLW and PLLW, the glm function in base R was used, with the brglm2 model used for PLB given one group did not lose any lambs between scanning and birth. In addition to group, other fixed effects included in the model for all maternal traits were the age of the ewe, the weight of the ewe prior to joining with the ram, and the average birth weight of her lambs. In addition to group, other fixed effects included in the model for LSW were age of dam, number of lambs born and sex of the lamb with dam being fit as a random effect. For BWt, other fixed effects included bdev, number of lambs born, weight of ewe just prior to joining, sex of the lamb with dam being fit as a random effect. The model for WWt included all factors fitted for BWt as well as the average birth weight of all lambs from its dam.

PAC data was processed as described (Jonker et al., 2018) and analysed using linear models. For analysis of methane production, the fixed effects in the model included age of the ewe, time off pasture, CO2 concentrations, number of lambs scanned and foetal age. For analysis of carbon dioxide measurement, the fixed effects in the model included age of the ewe, time off pasture, liveweight at time of PAC measurement, number of lambs scanned and foetal age.

Results

There was no evidence of an effect of PAC measurements on any of the ewe traits analysed ($P>0.50$; Table 2). Considering all lamb mortality from scanning to weaning (PLW), lamb survival was on average greater than 80% for all groups (Table 2). Little foetal loss was noted between pregnancy scanning and time of birth (PLB) with no factors included in the model being shown to affect this trait ($P>0.95$). Between scanning and birth, including losses during birth (PLBA), showed that mortality, was less than 8% for all groups. Loss between birth and weaning, considering only lambs that were born alive (PLLW), was less than 12% for all groups and no factors included in the model affected this trait ($P>0.05$). There was no evidence of a negative effect of PAC measurement on any of the lamb traits analysed (Table 2).

Table 1: Mean (±sem) age, number of foetuses at mid-pregnancy (NFMP), the weight of the ewe at joining of the ewes and rams (Ewe Wt), foetal age at time of first PAC measurement (foetal age) and the deviation (in days) in the birthdate relative to the mean birthdate of all lambs (bdev) are presented.

	Group				
Variable	Control	PAC mid-pregnancy	PAC later pregnancy	Feed restriction	P-value
Age	3.82 ± 0.91	3.82 ± 0.93	3.77 ± 0.87	3.76 ± 0.88	0.988
NFMP	2.03 ± 0.54	1.95 ± 0.57	2.03 ± 0.63	1.97 ± 0.54	0.909
Ewe Wt	65 ± 6.74	65.4 ± 6.33	63.8 ± 6.59	64.4 ± 5.25	0.705
Foetal age	69.8 ± 3.75	69.4 ± 3.82	69.6 ± 3.89	69.4 ± 4.02	0.953
bdev	-0.17 ± 4.12	0.12 ± 4.58	-0.01 ± 4.33	0.07 ± 4.61	0.993

Survival of lambs that were alive at birth to weaning was greater than 90% for all groups and no factor included in the model was shown to affect this trait ($P>0.05$). Average birth weight of the lambs ranged from 4.8 to 5.1 kg. While PAC measurement did not affect birth weight, all other factors included in the model had a significant effect ($P<0.05$). Average weaning weight ranged from 28.1 to 29.3 kg. While PAC measurement did not have a significant effect on weaning weight, and ewe weight at joining only approached significance ($P<0.10$), all other factors included in the model significantly affected weaning weight ($P<0.01$).

Methane concentrations were calculated for the ewes around day 100 of pregnancy and the average methane production was 23.2 ± 5.8 whereas carbon dioxide averaged 1390.7 ± 281.8 g per day with an average time off feed of 2.64 hours. With the limited sample size available, no differences ($P>0.15$) among ewes with different litter sizes and no effects of gestational age were observed for either methane or carbon dioxide measurements.

Discussion

The current study provides evidence that methane measurement using PAC around mid-gestation or approximately 6 weeks prior to birth was not detrimental to either the ewe or the lamb. Given the restricted air exchange during the PAC measurement, there was concern that reduced oxygen availability might affect the pregnancy/lambs. Severe hypoxia late in gestation (day 131) through occlusion of the umbilical cord for just 10 minutes has been reported to result in preterm labour and delivery in ewes (Baburamani et al. 2021). Foetal blood oxygen levels are correlated with weight of the foetus during near term (day 139-142 of gestation) (Flouri et al. 2022). Previous work has shown that long term hypoxia during later pregnancy, such as that caused by undernutrition and/or carrying of multiple foetuses, was associated with reduced foetal weights (Sales et al. 2018).

However, there was no evidence that any ewe aborted their pregnancy following PAC measurements and partial loss of multiple pregnancies was very low and not different among groups. Similarly, birth date deviation did not differ between groups indicating that gestational length was similar in the groups of ewes. The birth weight of the lambs was not reduced by exposure to the PAC and their growth patterns were normal. Lamb loss was similar among groups throughout the period of mid-gestation through to weaning. Taken together, this provides evidence that the lambs born following measurement of methane in the PAC were not affected by the potential limitation of oxygen supply. Concentrations of methane and carbon dioxide produced per day was roughly similar, albeit that methane concentrations appear slightly greater to what is observed in non-pregnant ewes, (Jonker et al. 2018; Johnson et al. 2022) with no

Table 2. The table of means and 95 % CIs for each trait for each group. (See text for definition of abbreviations).

	Group [Means (CI)]				
Trait	Control	PAC mid-pregnancy	PAC later pregnancy	Feed restriction	P-value (Type III)
PLW	0.898 (0.805, 0.949)	0.865 (0.762, 0.928)	0.857 (0.753, 0.921)	0.817 (0.704, 0.893)	0.553
PLB	0.994 (0.927, 1.000)	0.979 (0.904, 0.996)	0.986 (0.922, 0.998)	0.966 (0.887, 0.991)	0.963
PLBA	0.932 (0.848, 0.971)	0.928 (0.841, 0.969)	0.946 (0.868, 0.979)	0.925 (0.836, 0.968)	0.945
PLLW	0.951 (0.866, 0.983)	0.914 (0.812, 0.963)	0.904 (0.804, 0.955)	0.884 (0.778, 0.944)	0.526
LSW	0.942 (0.706, 0.991)	0.941 (0.712, 0.991)	0.908 (0.667, 0.980)	0.917 (0.656, 0.985)	0.874
BWt	5.0 (4.69, 5.36)	4.8 (4.48, 5.16)	5.1 (4.77, 5.39)	5.0 (4.62 ,5.28)	0.285
WWt	28.1 (26.18, 29.92)	29.0 (27.10, 30.93)	29.1 (27.44, 30.83)	29.3 (27.40, 31.15)	0.378

observable effect of number of foetuses the ewe was carrying or foetal age. However, it should be noted that the number of animals in the current study was relatively low with a limited gestational range and further research to better understand the effects of differing litter size and gestational age would be warranted.

Measurement of methane using the portable accumulation chambers during mid-pregnancy up to approximately day 110 of gestation does not appear to negatively affect pregnancy outcome for the ewe or lamb survival or growth. PAC provides a useful tool to better understand methane production of the ewe throughout the year including during pregnancy. This can provide a tool to obtain a better understanding of the interaction of pregnancy with methane production as well as any methane reduction strategies. It should however be noted that this trial contained a relatively small number of animals and continued monitoring of the effects of use of PAC in larger numbers of pregnant females would be prudent.

Acknowledgements

This work was supported by AgResearch's Strategic Science Investment Fund from NZ Ministry of Business, Innovation & Employment. Approval for publication was granted by AgResearch PreSeed (KiwiNet). Thank-you to the AgResearch Invermay farm team and the research technicians for all their help during the trial.

References

Baburamani AA, Tran NT, Castillo-Melendez M, Yawno MT, Walker DW. 2021. Brief hypoxia in late gestation sheep causes prolonged disruption of fetal electrographic, breathing behaviours and can result in early labour. Journal of Physiology 599:3221-3236. doi: 10.1113/JP281266

Beauchemin KA, Ungerfeld EM, Eckard RJ, Wang M. 2020. Review: Fifty years of research on rumen methanogenesis: lessons learned and future challenges for mitigation. Animal 14:s2-s16. doi: 10.1017/S1751731119003100

Flouri, D, Darby JRT, Holman SL, Cho SKS, Dimasi CG, Perumal SR, Ourselin S, Aughwane R, Mufti N, Macgowan CK, Seed M, David AL, Melbourne A, Morrison JL. 2022. Placental MRI Predicts Fetal Oxygenation and Growth Rates in Sheep and Human Pregnancy. Advanced Science (Weinh) 9(30):e2203738. doi: 10.1002/advs.202203738

Goopy JP, Woodgate R, Donaldson A, Robinson DL, Hegarty RS. 2011. Validation of a short-term methane measurement using portable static chambers to estimate daily methane production in sheep. Animal Feed Science and Technology 166-167:219-226.

Johnson PL, Hickey S, Knowler K, Wing J, Bryson B, Hall M, Jonker A, Janssen PH, Dodds KG, McEwan JC, Rowe SJ. 2022. Genetic parameters for residual feed intake, methane emissions, and body composition in New Zealand maternal sheep. Frontiers in genetics 13:911639. doi: 10.3389/fgene.2022.911639

Jonker A, Hickey SM, Rowe SJ, Janssen PH, Shackell GH, Elmes S, Bain WE, Wing J, Greer GJ, Bryson B, MacLean S, Dodds KG, Pinares-Patino CS, Young EA, Knowler K, Pickering NK, McEwan JC. 2018. Genetic parameters of methane emissions determined using portable accumulation chambers in lambs and ewes grazing pasture and genetic correlations with emissions determined in respiration chambers. Journal of Animal Science 96(8):3031-3042. doi: 10.1093/jas/sky187

Lawrence C. 2022. Cool Sheep Programme. https://www.blnzgenetics.com/cool-sheep-programme [Accessed 5/03/2024 2024].

Ministry for the Environment 2023. New Zealand's Greenhouse Gas Inventory 1990-2021 Snapshot. https://environment.govt.nz/assets/publications/climate-change/New-Zealands-Greenhouse-Gas-Inventory-1990-2021-Snapshot.pdf [Accessed 8 Sept 2023].

Morrison JL. 2008. Sheep models of intrauterine growth restriction: fetal adaptations and consequences. Clinical and Experimental Pharmacology and Physiology 35(7):730-743. doi: 10.1111/j.1440-1681.2008.04975.x

Sales F, Peralta OA, Narbona E, McCoard S, De Los Reyes M, Gonzalez-Bulnes A, Parraguez VH. 2018. Hypoxia and Oxidative Stress Are Associated with Reduced Fetal Growth in Twin and Undernourished Sheep Pregnancies. Animals (Basel) 8(11)doi: 10.3390/ani8110217

New Zealand Journal of Animal Science and Production 2024 Vol 84. 11-14

BRIEF COMMUNICATION: The behaviour and impact of sheep accessing a natural waterway in summer

A Bunyaga[1,2,*], C Barnes, I Draganova [2], P Kenyon [2], L Burkitt [2], & R Corner-Thomas [2]

[1] College of Veterinary Medicine and Biomedical Sciences, Sokoine University of Agriculture, P. O. Box 3020, Morogoro, Tanzania.
[2] School of Agriculture and Environment, Massey University, Private bag 11222, Palmerston North, 4442, New Zealand.
*Corresponding author email: asbunya@gmail.com

Introduction

In New Zealand (NZ) during summer sheep likely have a greater need to drink water than in other seasons due to higher dry matter content of pasture and warmer environmental conditions. On many farms in NZ water is provided to sheep from reticulated water troughs as well as natural sources such as creeks, streams and dams. In hot environments sheep show a preference for grazing close to water sources (Lynch 1972; Turner 1979). Thus, in summer in NZ sheep may have a greater motivation to access natural waterways to drink than in other seasons. While in winter, ewes have shown minimal interaction a stream and access to a water trough, or not, had no effect on ewe behaviour within 3m of a stream (Bunyaga et al. 2023).

Farming activities in NZ can potentially affect water quality due to contamination of waterways with sediment and the deposition of faecal matter (Ledgard et al. 1996; Monaghan et al. 2005). Indeed, cattle have been shown to have a significant impact on water quality of natural waterways (Cournane et al. 2011; O'Callaghan et al. 2019). Cattle are natural swimmers and will cross waterways to access grazing areas whereas sheep generally avoid entering waterways voluntarily (Ekesbo and Gunnarsson 2018). There is sparse information on the behaviour of sheep around natural waterways and their impact on water quality in New Zealand.

The aim of the current study was to examine ewe behaviour and interaction with a natural waterway during summer.

Materials and methods

Procedures in this study were carried out with the approval of the Massey University Animal Ethics Committee (MUAEC 19/62). The study was conducted over two weeks from the 15th of February (D1) to the 28th of February 2020 (D14; Table 1) at Massey University's Tuapaka farm, 15 km north-east of Palmerston North, New Zealand (40.3345° S, 175.7390° E.).

A crossover study was conducted with 40 ewes grazed in a single paddock that contained a discrete natural stream which was 233m in length, <1 m wide and <30 cm deep for two weeks (Figure 1). Ewes were offered access to a reticulated water trough (unrestricted) in the first week which was then covered in the second week to prevent access (restricted). Ewe movement within the paddock and interactions with the waterway was monitored using GPS, accelerometer and video surveillance footage as per Bunyaga et al. (2023). Pasture moisture was determined from pasture grab samples collected on study days 1, 8 and 13 (D1, D8 and D13). Weather data including rainfall (mm), relative humidity (%), air temperature (⁰C), solar radiation (MJ/mÂ²) and wind speed (m/s) was downloaded from a weather station located 800m from the study site.

GPS data were analysed using mapping software (ArcGIS Pro 2.2.4, 2018). An optimized hot spot analysis (z-score) was conducted to identify statistically significant spatial clustering of ewe GPS location fixes. Ewe behaviour data were checked for normality. Behavioural data including proportion of time ewes spent grazing, drinking, and walking when in the stream zone (within 3 m of the stream) were analysed using a 2-factor analysis of variance. A linear regression was also used to determine if any ewe behaviours were associated with time of the day or environmental temperature.

Results

Rainfall was recorded on 4 days of the two-week study period, with volumes ranging from 0.6 to 13.8 mm/day (Fig. 1). Maximum daily temperatures ranged from 17.4 to 26.6 °C, and the minimum temperature from 10.0 to 18.6 °C. Relative humidity ranged from 63.9% to 87.9%.

Animal density and spatial distribution

When ewes were restricted from accessing the water trough the optimized hot spot analysis identified one statistically significant spatial clustering of ewe GPS location fixes (hot spot; $p<0.05$) in the eastern area of the paddock which contained 46% of all fixes (Fig. 2A). When trough access was unrestricted, however, there was an additional hot spot in the north area near the trough which, when combined with the east hotspot, contained 61% of all the fixes.

The stream zone constituted 9% of the entire paddock area and contained 4.3 and 4.6% of all ewe GPS fixes during periods when the trough was restricted and unrestricted, respectively. When the stream zone was analysed in isolation there were seven and six hot spots

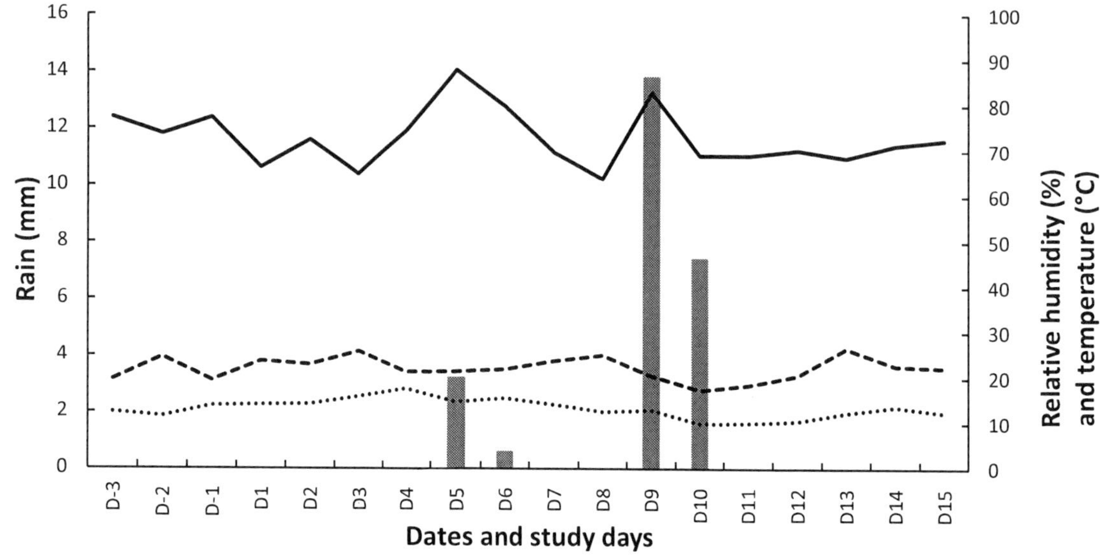

Figure 1: Daily mean rainfall (mm, bar), relative humidity (% solid line), minimum temperature (°C dotted line) and maximum temperature (°C, dashed line) during the study period. D-3 to D15 indicate number of days relative to the start of the study (D1;15 Feb 2020).

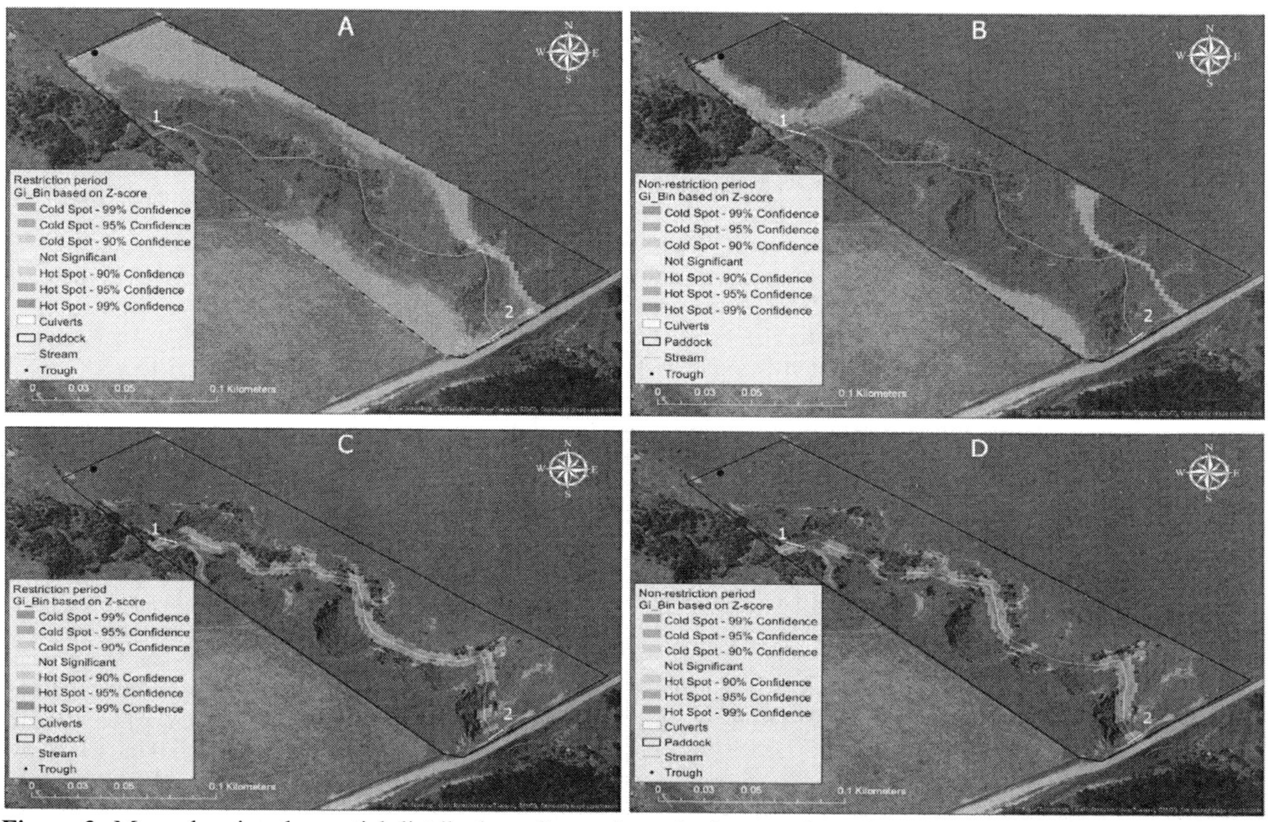

Figure 2: Maps showing the spatial distribution of ewes (magnitude per unit area) study paddock (A and B) and stream zone (within 3m, C and D) when access to the water-trough was restricted (panel A and C) or unrestricted (panel B and D) using optimised hot spot analysis. Blue areas indicate spatial clusters of low values ($p<0.05$; smaller negative z-score) representing low ewe density. Red areas indicate spatial clusters of high values ($p<0.05$; larger positive z-score) or high ewe density. White indicates random distribution or no spatial clustering.

containing 40 and 42% of fixes during the restricted and unrestricted periods, respectively ($p<0.05$; Fig. 2D). There were three cold spots (low ewe density) in the stream zones in both treatment periods.

Time spent in the stream zone

The duration that ewes were detected within 3m of the stream differed by time-of-day ($p<0.05$), with greater durations during day time periods followed by evening, and night (Table 6). Time spent within 3m of the stream did not differ ($p=0.81$) between the period the trough was restricted (13.3 ± 4.4 min/ewe/day) and unrestricted (10.3 ± 2.0 min/ewe/day).

Table 1: The daily duration (min/ewe/day mean ± SEM) that each ewe spent within 3m of any camera location by time of day during the entire study

Time of day *	n	Duration (min/ewe/day)
Night	126	6.6 ± 1.11[a]
Morning	84	11.8 ± 0.90[b]
Day	70	22.7 ± 1.02[d]
Evening	56	17.3 ± 1.38[c]

[abcd] superscripts with different letters are significantly different ($p<0.05$). * Night = 2000 to 0500 h, morning = 0600 to 0800 h, daylight = 0900 to 1600 h and evening = 1700 to 1900 h.

Behaviour within the stream zone

Observations from video footage over the 14 days of the study showed that in the stream zone ewes spent 41.9% of the time grazing (n=2304 of 5496 occasions), 26.9% stationary (n=1476) and 21.0% walking (n=1152). Ewes were observed to sniff and drink from the stream on 72 (1.3%) and 396 (7.2%) individual ewe occasions, respectively. On twelve individual occasions ewes were observed to walk in the stream (0.2%). Ewes spent more time grazing and drinking when the water trough was restricted than when access was provided ($p<0.05$; Fig. 3). The frequency of stationary and walking behaviours were unaffected by treatment ($p>0.05$).

Discussion

The aim of the current study was to examine the behaviour of sheep around a natural waterway in summer when access to a water trough was provided or unrestricted. It was hypothesised that in the summer, in the Manawatū region of NZ, access to a reticulated water trough would influence the frequency of ewe interactions with the natural waterway.

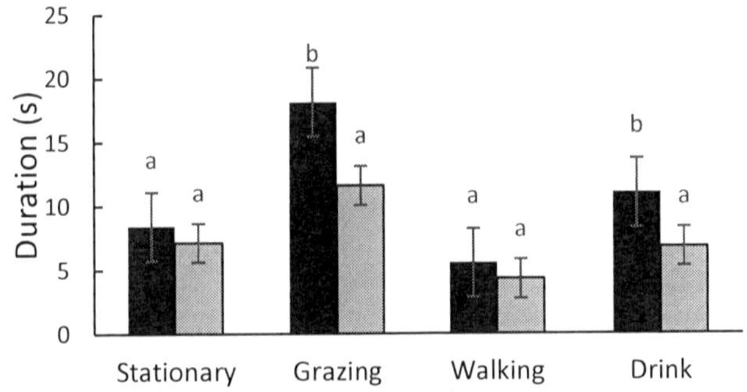

Figure 3: The average duration (time in seconds ± SE) ewes were observed to be stationary, grazing, walking, or drinking in the stream zone during periods of restricted access to water trough (black bars) or when access was not restricted (grey bars). Within each behavioural event, means with different letters are significantly different ($p<0.05$)

When in the stream zone, ewes were observed to graze and drink more frequently during the period when the water trough was restricted compared to when access was provided. The time grazing in the stream zone, regardless of trough access, was similar to the 49.5% of time grazing reported by Filipčík et al. (2020). Ewes in the current study spent 7.2% of their time when within the stream zone drinking from the stream. Observations of the behaviour of extensively managed sheep in West Africa during day light hours reported that in 4.4% and 3.4% of observations sheep were drinking in the dry and cool seasons, respectively (Ouédraogo-Koné et al. 2006). This difference in findings was likely due to the opportunistic recording of ewes when near the stream utilised in the current study rather than continuous monitoring.

When ewes had access to the water trough in the current study there was a significant clustering of ewes (hotspot) close to the trough that was not seen during the restricted period. This suggests that when the water trough was available the ewes utilised it. In winter, when access to the trough was unrestricted, there was no clustering of ewes near the trough (Bunyaga et al., 2023). Markwick (2007) reported that the consumption of free water by sheep was 40% greater in summer than winter. This increase in water consumption is likely to be a combination of warmer environmental conditions and the lower moisture content of pasture. The moisture content of the pasture offered in the current study was 56% compared with 77% in winter (Bunyaga et al. 2023). This likely necessitated the need for sheep to seek water. (Macfarlane et al. 1966; MacFarlane and Howard 1972) demonstrated that when pasture contained more than 60% water the water requirements of sheep could be met through forage consumed alone.

In the current study the duration ewes spent within 3m of the stream was greater during the hours of daylight (morning and day) than in the evening, with the least at night. It was expected that fewer ewes would be observed during the day as sheep tend to be inactive or avoid grazing during the middle of the day (Shinde et al. 1997; Evangelou et al. 2010). This finding was contrary to previous studies which reported that most of activity was during midday and evening (McGranahan et al. 2018; Filipčík et al. 2020).

In conclusion, sheep spent proportionally less time in the stream zone than would be expected based on its area. Ewes were observed to interact directly with the waterway, although, this was not influenced by the availability of the reticulated water trough. Further long-term studies are required to verify these results when ambient temperatures are greater than experienced in the current study and across varying environments.

References

Bunyaga A, Corner-Thomas R, Draganova I, Kenyon P, Burkitt L. 2023. The Behaviour of Sheep around a Natural Waterway and Impact on Water Quality during Winter in New Zealand. Animals. 13(9): 1461.

Cournane FC, McDowell R, Littlejohn R, Condron L. 2011. Effects of cattle, sheep and deer grazing on soil physical quality and losses of phosphorus and suspended sediment losses in surface runoff. Agriculture, Ecosystems & Environment. 140(1): 264–272.

Ekesbo I, Gunnarsson S. 2018. Farm animal behaviour: characteristics for assessment of health and welfare. Wallingford, UK: CABI. 341 p

Evangelou CH, Yiakoulaki MD, Papanastasis VP. 2010. Seasonal changes in grazing activities of sheep and goats in communal mediterranean rangelands of northern Greece. In: Proceedings of the 7th National Conference on Rangelands Xanthi Greece. Pg. 249–253.

Filipčík R, Hanzlikova K, Kuchtik J, Reckova Z, Hosek M, Pesan V, Kopec T. 2020. Assessment of some behavioral patterns of a sheep flock on pasture. Animal Welfare, Etológia és Tartástechnológia. 16(1):8–14.

Ledgard SF, Thom ER, Singleton PL, Thorrold BS, Edmeades DC. 1996. Environmental impacts of dairy systems. In: Proceedings 48th Ruakura Farmers' Conference. AgResearch Hamilton, New Zealand.

Lynch JJ. 1972. Movement of some rangeland herbivores in relation to their feed and water supply. In: The Impact of Herbivores on Arid and Semi-arid Rangelands Proceedings of the 2nd United States/Australia Rangeland Panel, Adelaide. Pg. 163–72.

Macfarlane WV, Dolling CHS, Howard B. 1966. Distribution and turnover of water in Merino sheep selected for high wool production. Australian Journal of Agricultural Research. 17(4):491–502.

MacFarlane WV, Howard B. 1972. Comparative water and energy economy of wild and domestic mammals. In: Proceedings of the Symposium of the Zoological Society of London. 31: 261–296.

Markwick G. 2007. Water requirements for sheep and cattle. Profitable & sustainable primary industry. 20. Retrieved from https://www.boprc.govt.nz/media/470831/reasonable-stock-water-requirements-guidelines-horizons.pdf

McGranahan DA, Geaumont B, Spiess JW. 2018. Assessment of a livestock GPS collar based on an open-source datalogger informs best practices for logging intensity. Ecology and Evolution. 8(11):5649–5660.

Monaghan RM, Paton RJ, Smith LC, Drewry JJ, Littlejohn RP. 2005. The impacts of nitrogen fertilisation and increased stocking rate on pasture yield, soil physical condition and nutrient losses in drainage from a cattle-grazed pasture. New Zealand Journal of Agricultural Research. 48(2):227–240.

O'Callaghan P, Kelly-Quinn M, Jennings E, Antunes P, O'Sullivan M, Fenton O, hUallacháin DÓ. 2019. The Environmental Impact of Cattle Access to Watercourses: A Review. Journal of Environmental Quality. 48(2):340–351.

Ouédraogo-Koné S, Kaboré-Zoungrana CY, Ledin I. 2006. Behaviour of goats, sheep and cattle on natural pasture in the sub-humid zone of West Africa. Livestock Science. 105(1):244–252.

Shinde AK, Karim SA, Patnayak BC, Mann JS. 1997. Dietary preference and grazing behaviour of sheep on Cenchrus ciliaris pasture in a semi-arid region of India. Small Ruminant Research. 26(1):119–122.

Turner JC. 1979. Water, energy and electrolyte, balance in the desert bighorn sheep, Ovis canadensis. Riverside, CA, USA: University of California.

New Zealand Journal of Animal Science and Production 2024 Vol 84: 15-18

BRIEF COMMUNICATION: Influence of the 'SLICK' allele of the prolactin receptor gene (PRLR) on the growth and behaviour of dairy heifers during winter.

LA Robb [1]*, RH Bryant[1], BP Beckett[1], EG Donkersloot[2], GM Worth[2], Huitong Zhou[1], and JGH. Hickford[1]
[1]*Faculty of Agriculture and Life Sciences, Lincoln University, PO Box 85084, Lincoln 7647, Christchurch, New Zealand.*
[2]*Livestock Improvement Corporation Ltd., Newstead, Hamilton 3240, New Zealand*
**Corresponding author.* Email: Luke.Robb@lincolnuni.ac.nz

Abstract

This study investigated the effects of the SLICK allele on the responses of dairy heifers in cold conditions in New Zealand. Three groups (n=9/group), control, non-SLICK and SLICK progeny of bulls heterozygous for the SLICK allele were investigated. These heifers were grazed together to identify their responses to winter conditions under adequate, followed by restricted feed allocation. The control group had an 87% lower growth rate during the restricted period compared to the SLICK group (0.064 vs 0.484 kg/d, P = 0.035). During restricted feeding conditions, the SLICK heifers showed improved pasture harvesting efficiency, by maintaining growth rate while having a lower eating time, compared to the control heifers (277 vs 317 ± 10.9 mins/day respectively, P = 0.017). The SLICK group were also less active than the control group. These differences indicate more research is needed to understand the wider role of the prolactin axis on the physiological responses of SLICK cattle.

Keywords: Heat stress; Cold challenge; Dairy cattle; SLICK allele; Prolactin receptor gene.

Introduction

Despite a temperate climate, New Zealand will be affected by climate change and an increased potential for extreme seasonal weather events, with the number of tropical nights (defined as when the daily temperature does not fall below 20°C) to increase an average >1000% nationally and >2000% in regions like Canterbury and Taranaki from 2024 onwards (Abercrombie et al. 2018). Breeding for improved heat tolerance in livestock is one approach to ensure future welfare and productive performance under warming conditions. In 2014, a genetic variant known as the SLICK allele was introduced into Holstein-Friesian Jersey crossbred cattle to ascertain its relevance in a temperate climate dairy industry. The SLICK allele is caused by a single base pair deletion in the prolactin receptor gene (*PRLR*), originating in Senepol cattle (Littlejohn et al. 2014). When investigated under New Zealand heat load conditions, SLICK carrier Holstein-Friesian cattle displayed improved thermoregulation (Worth et al. 2023).

The short, sleek coat of SLICK cattle has been described as an important contributor to their heat tolerance (Dikmen et al. 2014), but little is known about whether SLICK carrier cattle can or need to adapt their behaviour and/or physiology, when facing a cold challenge. The phenotypic advantage in periods of heat challenge may therefore potentially be disadvantageous in the winter. This would be of concern for New Zealand's typically outdoor cattle systems, where only 85-89% of heifer replacements reach target live weight by their first calving (McNaughton and Lopdell, 2013). Accordingly, the purpose of this research was to investigate if the presence of the SLICK allele influenced heifer growth rates, behaviour, or other dairy cattle traits in winter conditions.

Materials and Methods

Experimental site and design

The study was conducted with approval from the Lincoln University Animal Ethics Committee (#2023-17) at the Lincoln University Research Dairy Farm (-43.64, 172.46) between May and July 2023. The experiment was a randomised design with three treatment groups, and nine heifers per treatment. The control heifers were the progeny of AI sires selected for high breeding worth, while the non-SLICK group consisted of heifers born to AI sires from LIC that were heterozygous for the SLICK allele, but did not inherit the SLICK allele, and the SLICK heifers were the progeny of the same carrier bulls carrying the SLICK allele. All heifers were approximately 10 months of age and had an average breeding worth (daughter proven) of $234 ± 27.4, $125 ± 67.7 and $125 ± 95.5 (mean ± sd) for the control, non-SLICK and SLICK respectively.

The absence or presence of the SLICK allele in the heifers was confirmed using DNA extracted from ear blood samples (Zhou et al., 2006) using a polymerase chain reaction-single strand conformation polymorphism approach.

All the heifers had been under the same management from birth and had been reared together on pasture since weaning. On 13 June 2023, the treatment heifers were drafted from a larger group of heifers and allocated 28 m^2/animal/day of perennial ryegrass and white clover pasture using temporary fencing. During the experimental period, supplement was fed as pasture baleage at 1 kg DM/head/day. Between 13 and 27 June was the unrestricted feed period where all heifers were offered approximately 3.5% of bodyweight as pasture (above a 2cm residual). This was reduced to an allocation of 2.8% of bodyweight between 27 June and 11 July as the restricted feeding period. The restricted feed period was used to reduce the energy available for metabolic heat production to increase the physiological factors which lead to cold stress, in response to the milder climatic conditions than anticipated. The allocation area and supplement remained the same during the entire observation, but during the restricted period the pre-graze pasture mass reduced from 2550 ± 194.8 to 2080 ± 69.9 kg DM/ha (± SEM).

Measurements

Live weight was measured (un-fasted) on four occasions: a covariate weight taken on 15 May, and again when the heifers were drafted on 13 June, 27 June, and 11 July 2023. Behaviour was monitored between 22 June to 5 July, by attaching an ear mounted accelerometer (Senztag-NZ, Morrinsville, NZ) to six SLICK and five control heifers. The tri-axial accelerometer tags recorded minutes per hour for movement that is categorised as 'high activity', 'active', 'non-active', 'eating', and 'rumination' (Pereira et al. 2017).

Apparent dry matter intake was estimated using the pasture disappearance method using a calibrated rising plate meter (RPM, Jenquip, Feilding, New Zealand). Baleage dry matter content was determined from weekly subsamples of fed baleage, and refusals determined by visual estimation.

Climate data were collected from the NIWA CliFlo database, Broadfields weather station (approximately 2km from the experimental site; ID 17603). A temperature humidity index (THI) was calculated using the equation described by Davis et al. (2003):

$$THI = 0.8T + [RH * (T - 14.4)] + 46.4$$

where T is daily average temperature and RH is relative humidity.

Statistical analysis

The statistical analyses were conducted using GenStat (V. 22.1.0 VSN International Ltd). Growth rate, live weight, and behaviour (minutes/day per category) were compared using the general linear model procedure in GenStat. The heifer group and period (unrestricted and restricted) were included as fixed terms and animal as the random term. Pre-winter live weight on 15 May was included as a covariate. Behaviour data (minutes/hour) for the five categories of behaviour were analysed using repeated measures procedure in GenStat, with group and hour of the day (and their interaction) as fixed terms, and individual heifer as the random term.

The daily average temperature between 22 June and 11 July was 8.8 ± 0.89 °C and ranged between 4.1 and 13.4°C. The daily average temperature was 2.5 °C ($P = 0.011$) cooler than the 30-year average of 11.3 ± 0.25 °C (1992-2022). The average THI-values during the unrestricted and restricted periods were 49.8 ± 1.30 and 42.2 ± 1.70 °C respectively. According to Fu et al. (2022) cold stress is more likely to occur at THI < 38, and during the observation THI below 38 occurred on 2 out of a total of 14 days indicating limited cold challenge.

Pasture and supplement utilisation was 85 and 95% respectively resulting in a total daily apparent intake of 7.4 ± 0.48 and 5.8 ± 0.31 kg DM/heifer/day, in respective unrestricted and restricted periods. Restricting feed intake increases sensitivity to cold conditions by restricting the energy available for metabolic heat production (Wang et al. 2023). While we could not confirm if the heifers in the current experiment were experiencing cold stress, it is evident that they altered their behaviour (inactive and rumination time) in response to the change in nutritional and climatic conditions. Collectively the heifers (control and SLICK) increased rumination time from 230 to 290 ± 17.7 minutes/d between the unrestricted and restricted period ($P<0.05$). Eating time stayed the same between periods, but control heifers spent more time eating compared to SLICK (317 vs 277 ± 10.9 mins/d, $P < 0.05$). There was no difference between groups or periods for active (sum of active and high active) behaviour which was an average 437 mins/d. However, heifers spent more time inactive in the unrestricted compared with restricted period (460 vs 435 ± 7.4 mins/d respectively, $P < 0.05$) and SLICK heifers were more inactive than control heifers in both periods (468 vs 427 ± 7.4 mins/d respectively, $P = 0.001$, Table 1). Inactivity, in the form of recumbency, can be used as a heat and energy conservation method for young stock, primarily to reduce the surface area exposed to air. This reduces the amount of maintenance energy needed to maintain temperature homeostasis in cold conditions (Rawson et al. 1989).

Results and discussion

Table 1. Daily behaviour (minutes/day) of heifers comprised of CONTROL (n=6), and SLICK carrier (n=5) groups. Where SEM is the standard error of the mean.

	Non-restricted		Restricted					
	CONTROL	SLICK	CONTROL	SLICK	SEM	Group	Period	G x P
Eating	319	266	315	287	10.9	**0.017**	0.527	0.423
Ruminating	223	237	275	305	17.7	0.401	**0.025**	0.744
Active	203	208	188	158	14.6	0.536	0.113	0.405
High active	257	253	241	239	8.4	0.795	0.226	0.959
Not active	438	482	416	454	7.4	**0.001**	**0.029**	0.777

By the end of the study period live weight increased from 230 to 244 ± 1.0 kg between the unrestricted and restricted periods ($P < 0.001$), although there was no interaction with group for live weight ($P = 0.137$). The impact of cold and feed restriction on growth was a pronounced reduction in growth rate from 622 to

299 ± 56.4 g/d between the unrestricted and restricted period. There was a group-by-period interaction for growth rate ($P < 0.001$) in that the control heifers experienced much greater weight gain reductions than either of the other groups (Figure 1).

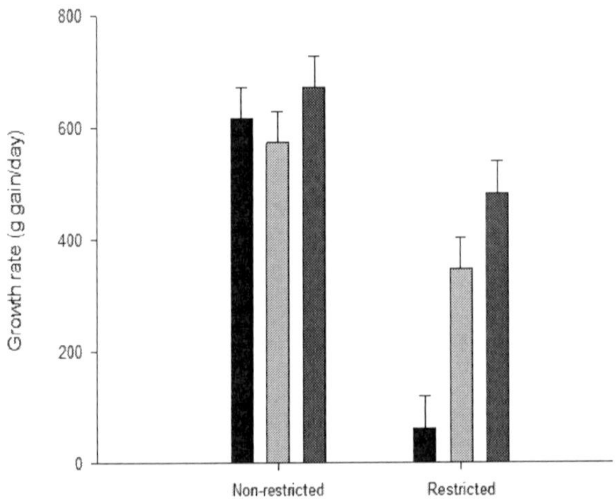

Figure 1. Growth rate (g/d) of heifers from different sires including control (black) versus non-SLICK (light grey) or SLICK (dark grey) and different feeding restriction during winter. Bars represent the standard error of the mean.

The dry matter intake requirements of cattle have been reported to be increased in periods of colder weather, compared to warmer periods (Wang et al. 2023), and based on the change in growth rate between the two periods, the control heifer's maintenance requirements increased (demonstrated by the 90% decrease in growth rate). In contrast, the SLICK heifers maintained (statistically) the same growth rate as in the unrestricted period, indicating that they were better at regulating heat loss as demonstrated by the higher inactive time.

Green et al. (2013) reported that the most efficient heifers spent less time eating per day compared to the least efficient heifers ($P = 0.026$). This would indicate that the SLICK heifers are able to be more efficient with the feed they eat and achieve this by modifying their grazing behaviour to spend more time eating during the restricted period. They were able to outcompete the control heifers for enough energy to maintain their growth rate.

We accept the small sample size and short duration of this study which limits our conclusions. However, as a pilot study these results indicate that SLICK heifers modified their behaviour to adapt to the lower feed allocation and conserve energy during cold conditions, which enabled them to outcompete control heifers. These results highlight the need for further investigation into behavioural adaptations to environment associated with breeding. The concept of behavioural genetics has previously been reported among beef animals (Garcia et al 2024) and our results indicate that SLICK heifers -in contrast to control - were able to maintain growth rates during a period of mild cold challenge and feed restriction. Further investigation into the behavioural differences of SLICK dairy animals, under a range of environmental conditions is warranted.

Acknowledgements

This work is part of a wider honours research with support from LIC's Patrick Shannon Scholarship 2023. We would like to thank the Lincoln University Research Dairy Farm staff for their assistance in undertaking the research on farm.

References

Abercrombie R, Sneddon N, Scarsbrook M, Arheimer B, Little L (2018). Predicting climate change impacts for dairy in New Zealand and Australia. Australian Dairy Science Symposium. researchgate.net/publication/329179282 (Accessed 18 September 2024)

Davis MS, Mader TL, Holt SM, Parkhurst AM 2003. Strategies to reduce feedlot cattle heat stress: Effects on tympanic temperature. Journal of Animal Science 81(3): 649-661. doi.org/10.2527/2003.813649x.

Dikmen SE, Khan FA, Huson HJ, Sonstegard TS, Moss JJ, Dahl GE, Hansen PJ 2014. The SLICK hair locus derived from Senepol cattle confers thermotolerance to intensively managed lactating Holstein cows. Journal of Dairy Science 97: 5508-5520. doi.org/10.3168/jds.2014-8087.

Fu X, Zhang Y, Zhang YG, Yin YL, Yan SC, Zhao YZ, Shen WZ 2022. Research and application of a new multilevel fuzzy comprehensive evaluation method for cold stress in dairy cows. Journal of Dairy Science 105 (11): 9137-9161. doi: 10.3168/jds.2022-21828.

Garcia C, Zhou H, Altimira D, Dynes R, Gregorini P, Jayathunga S, Maxwell T, Hickford J 2022. The glutamate metabotropic receptor 5 (GRM5) gene is associated with beef cattle home range and movement tortuosity. Journal of Animal Science and Biotechnology 13: 11. doi.org/10.1186/s40104-022-00755-7

Green TC, Jago JG, Macdonald KA, Waghorn GC 2013. Relationships between residual feed intake, average daily gain, and feeding behavior in growing dairy heifers. Journal of Dairy Science 96 (5): 3098-3107. doi.org/10.3168/jds.2012-6087

Littlejohn MD, Henty KM, Tiplady K, Johnson T, Harland C, Lopdell T, Sherlock RG, Li W, Lukefahr SD, Shanks BC, Garrick DJ, Snell RG, Spelman RJ, Davis SR 2014. Functionally reciprocal mutations of the prolactin signalling pathway define hairy and slick cattle. Nature

Communications 5: 5861. doi.org/10.1038/ncomms6861

McNaughton LR, Lopdell TJ 2013. Effect of heifer liveweight on calving pattern and milk production. Proceedings of the New Zealand Society of Animal Production 73: 103–107.

Pereira GM, Heins BJ, Endre MI 2018. Technical note: Validation of an ear-tag accelerometer sensor to determine rumination, eating, and activity behaviours of grazing dairy cattle. Journal of Dairy Science 101(3): 2492-2495. doi.org/10.3168/jds.2016-12534.

Rawson R, Dziuk H, Good A, Anderson J, Bates D, Ruth G, Serfass R 1989. Health and Metabolic Responses of Young Calves Housed at -30 C to -8°C. Canadian Journal of Veterinary Research 53: 268-274. https://www.ncbi.nlm.nih.gov/pmc/articles/PMC1255709/pdf/cjvetres00051-0016.pdf (accessed 26 April 2024)

Wang S, Li Q, Peng J, Niu H 2023. Effects of long-term cold stress on growth performance, behaviour, physiological parameters, and energy metabolism in growing beef cattle. Animals 13(10): 1619. doi.org/10.3390/ani13101619

Worth G, Donkersloot E, Spelman R, Davis S 2023. BRIEF COMMUNICATION: Effects of the SLICK gene on heat tolerance in grazing dairy cattle. New Zealand Journal of Animal Science and Production 83: 5-7.

Zhou H, Hickford JGH, Fang Q 2006. A two-step procedure for extracting genomic DNA from dried blood spots on filter paper for polymerase chain reaction amplification. Analytical Biochemistry 354(1): 159-161. doi: 10.1016/j.ab.2006.03.042.

New Zealand Journal of Animal Science and Production 2024 Vol 84. 19-22

BRIEF COMMUNICATION: Exploration of the New Zealand Deer Rumen Microbiome

JL Clarke[1,2]*, TP Bilton[1], H Henry[1], KM McRae[1], F Booker[1], B Veenvliet[1], W Bain [1], JA Ward[1], B Thompson[1], JC McEwan[1], BJ Perry[1], MA Black[2], SJ Rowe[1]

[1] *AgResearch Invermay Agricultural Centre, Private Bag 50034, Mosgiel, 9053, New Zealand*
[2] *Department of Biochemistry, University of Otago, Dunedin, 9016, New Zealand.*
*Corresponding Author. jordan.clarke@agresearch.co.nz

Abstract
Rumen microbial communities (RMC) have been shown to be affected by the host individual's genetics and are heritable in sheep and cattle. RMC are predictive of feed related traits such as methane emissions, a by-product of the fermentation process. The aim of this project was to determine the variability in RMC amongst deer and to estimate heritability of the deer RMC. The rumen contents of 409 red and wapiti cross deer were sampled, and microbial DNA was sequenced. The genome taxonomy database was used to determine taxonomy assignment at the genus level. Sequences were used to determine similarity of RMC amongst individuals and presence of known microbes. Similar to other ruminants, *Prevotella* was the most abundant genus present in the deer RMC. Dimension reduction methods were applied to three standardisation methods to examine the variation in the RMC. Heritability estimates for the first two components of the reduced dimensions of the RMC ranged from 0.078 (± 0.058) to 0.72 (± 0.21) across the standardisation methods. This suggests that the deer RMC is controlled by host genomics.

Keywords; Deer, Rumen, Microbiome, Heritability

Introduction

Grazing ruminants can digest complex carbohydrates such as plant cell walls through a symbiotic relationship with microbes in the rumen. The fermentation pathways, along with the resulting energy sources and waste products, vary depending on the rumen microbial communities (RMC) present in the gut (Keum et al., 2024). A prevalent byproduct of this fermentation is methane, which can be predicted based on diet composition and the bacteria that drive fermentation pathways, producing varying amounts of hydrogen substrate (Clemmons et al., 2019, Rabee et al., 2024). Mitigation of methane emissions from ruminants can be achieved through selective breeding, where a key factor is the prediction of RMC (Hess et al., 2022). Although diet significantly influences RMC, they have been shown to be heritable in ruminants like sheep and cattle, with heritability estimates ranging from 0.1 to 0.4 (González-Recio et al., 2023). Therefore, RMC can be considered a phenotype for inclusion in selection. This has been done successfully in sheep, with the combination of host genetics and metagenomic profiles (Hess et al., 2022). Whilst the deer RMC have not been as intensively explored as other ruminants, it has been associated with disease immunity, resistance to pathogenic bacteria, and feed efficiency (Hu et al., 2017). Hence, the deer RMC may also be heritable and be used to increase prediction accuracy of breeding values in deer.

Deer are highly adaptable and versatile animals, thriving in a broad range of environments (Berlioz et al., 2017). The RMC of sheep and cattle have been shown to have similar heritability and be predictive of methane. Deer are browsers and adapt to seasons, therefore have different digestive characteristics than other ruminants. Hence, further investigations into deer RMC are crucial to expanding the industry and reducing methane emissions. The aim of this paper is to investigate variations in New Zealand deer RMC and estimate the heritability of the RMC.

Materials and Methods

Experimental animals and protocols applied in this study were approved by the AgResearch Ruakura (Hamilton, NZ) Animal Ethics committee (approvals 14906, 14907 and 14981).

Animals & phenotypes.

The rumen contents of 409 red and wapiti cross deer (*Cervus elaphus* and *Cervus canadensis* respectively) from three different cohorts across New Zealand were sampled over three years. SNP data was generated for individuals using genotyping by sequencing (GBS) and sex, breed and year of birth (YOB) were also recorded.

Rumen microbial counts.

Microbial DNA was sequenced using restriction enzyme reduced representation sequencing (Hess et al., 2020). Sequences were processed through the reference-based pipeline described in Hess et al. (2020) using the genome taxonomy database (GTDB) (Parks et al., 2021) to classify microbial sequences at the genera level. The microbial count matrix was created using relative abundance of genera per sequence, and filtered so that only genera that had a relative abundance greater than 0.01% within at least one cohort were retained.

Statistical models.

Three standardisation methods were applied to the microbial count matrix. (1) Microbial counts were converted to a relative abundance matrix (i.e., counts were converted into proportions per sample), log10 transformed and scaled per genera within each cohort, then converted into a microbial relationship matrix (MRM) as described in Hess et al. (2020), (2) A centred-log ratio (CLR) transformation was applied to the microbial count matrix (creates a covariance matrix reducing the distance between counts to a normalised scale). (3) A Bray-Curtis dis-similarity matrix was generated using the CLR transformed matrix from (2) (measures dis-similarity between samples on a scale from 0-1). Different standardisation methods were used to ensure that heritability of the RMC was not an artifact of a particular standardisation method. Cohort correction ensured population stratification was removed. Principal Component Analyses (PCA) were applied to the MRM and CLR adjusted count matrices and Principal Coordinate Analyses (PCoA) were performed using the Bray-Curtis matrix as a distance matrix. The heritability of the first two components for each method were estimated using a linear mixed model with YOB, sex and breed fitted as fixed effects, and additive genetic effect using a genomic relationship matrix as variance/covariance matrix, where the response variable is the first or second component from the PCA or PCoA.

Results

The most prevalent genus in the deer RMC was *Prevotella* and the top ten most abundant genera make up ~50% of the total microbial counts across all individuals (Figure 1). Cohorts from 2022 and 2019 are the most similar, with the top three most abundant genera (*Prevotella*, *Limivicinus*, and *Cryptobacteroides*) making up ~25% of the total relative abundance. The 2021 cohort displays higher levels of *Prevotella* overall with a relative abundance ranging from ~19%-30% of the total microbial counts across individuals. There is microbial count variation across individuals, where individuals within the same cohort are the most similar.

The first component for the Hess standardisation (method 1) explains 28% of variation amongst the MRM, whilst component two explains 14%. Component one for the CLR standardisation (method 2) explained 18% of the variation across the RMC, whilst component two explains 12%. The heritability's estimated from the components ranged from 0.078-0.72 (Table 1). Component 1 displayed a higher heritability estimate than component two for methods 2 and 3. Component two was found to have a higher heritability estimate than component one in method 1. The standard errors for both components for method 3 were significantly lower than those found in methods 1 and 2. Component one of method 2 had the highest heritability estimate of 0.72 ± 0.21. Component two of method 3 had the lowest heritability estimate of 0.078 ± 0.058.

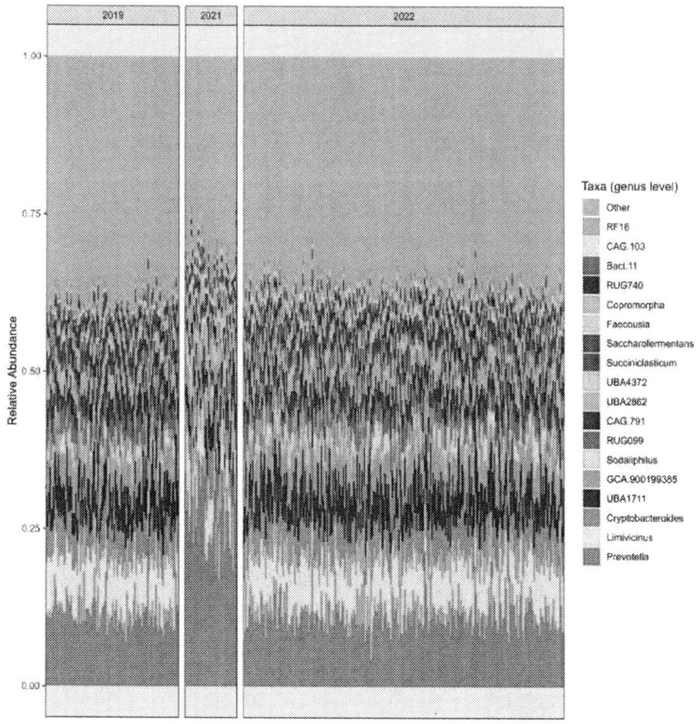

Figure 1 Relative abundance plot of rumen microbes by year of collection. Samples are placed across the x axis and relative abundance of genera on the y axis. The most abundant microbes are on the bottom, with decreasing abundances further up, where the total composition of genera adds to 1. Microbial genera are separated by colour and given on the right, most abundant on the bottom and moving upwards with least abundant.

Table 1. RMC heritability estimates per standardisation method. The response variable gives the components explaining the largest amount of variance in the model (1 and 2), with the proportion of variance explained in brackets.

Standardisation method	Response variable (component)	h^2	SE
Hess (1)[1]	1 (0.28)	0.17	0.14
	2 (0.14)	0.57	0.15
CLR (2)[2]	1 (0.18)	0.72	0.21
	2 (0.12)	0.44	0.25
Bray-Curtis (3)[3]	1 (-)	0.65	0.047
	2 (-)	0.078	0.058

[1]The Hess method (1) log corrects per cohort then scales per genera.

[2]The CLR method (2) creates a correlation matrix with CLR standardisation.

[3]The Bray-Curtis method (3) creates a dis-similarity matrix from the matrix in method 2.

Discussion

The most prevalent genera in the deer RMC was *Prevotella*, which is consistent with findings in cattle, goats, sheep and in farmed red and fallow deer (González-Recio et al., 2023, Wang et al., 2019, Rabee et al., 2024). In cattle *Prevotella* was observed to make up to 60% of the RMC, providing evidence of persistence across a variety of diets (Keum et al., 2024). *Prevotella* has been found to improve volatile fatty acid production which is a preferred energy source for ruminants (Rabee et al., 2024). *Prevotella* has been associated with higher methane production, however, due to being the dominant species, this does not inform much about manipulating methane output (Han et al., 2023). Another highly abundant genus was *Sodaliphilus*, which is associated with methane production in dairy calves, although the mechanism for this is largely unclear due to the ambiguity of its role in the rumen microbiome (Park et al., 2022). Therefore, the deer rumen microbiome shows similarities to that of other ruminants and may be used to predict methane emissions. The RMC was seen to vary across individuals, but the largest variation observed between cohorts. This is likely due to the individuals within cohort having the same environment and being in close contact (Wang et al., 2019). Because of this, the cohort effect has been corrected for heritability estimation.

The standardisation methods depict different aspects of the rumen microbial composition data due to the transformations undertaken. The Hess method (method 1) measured the relationships in the RMC between individuals adjusting for cohort variation. The CLR method with PCA (method 2) describes the most variable aspects of the RMC. The Bray-Curtis method (method 3) measures the dissimilarity in the RMC between individuals. The heritability estimates for the components for each model varied significantly ($0.078 \pm 0.058 - 0.72 \pm 0.21$). This is likely due to the first two components from the PCA or PCoA capturing different aspects of the RMC. However, since each method has a component with a significant non-zero heritability estimate, we conclude that aspects of the deer RMC is controlled by the host genome.

This study has successfully shown that components of the deer RMC are heritable. This knowledge may be used to improve breeding selection. Further research is required to investigate whether the RMC is associated with key production traits.

Acknowledgements

Funded by Deer Industry NZ, AgResearch and the University of Otago, Biochemistry.

References

Berlioz E, Azorit C, Blondel C, Sierra Tellado Ruiz M, Merceron G. 2017. Deer in an arid habitat: dental microwear textures track feeding adaptability. Hystrix, the Italian Journal of Mammalogy 28: 222-230.

Clemmons BA, Voy BH, Myer PR. 2019. Altering the Gut Microbiome of Cattle: Considerations of Host-Microbiome Interactions for Persistent Microbiome Manipulation. Microbial Ecology 77: 523-536.

González-Recio O, Martínez-Álvaro M, Tiezzi F, Saborío-Montero A, Maltecca C, Roehe R. 2023. Invited review: Novel methods and perspectives for modulating the rumen microbiome through selective breeding as a means to improve complex traits: Implications for methane emissions in cattle. Livestock Science 269: 105171.

Han Y, Li S, Mu R, Zhao F, Yan X, Si H, Li Z. 2023. Roe Deer Produce Less Methane and Harbor Distinct Gut Microbiota. Fermentation 9: 186.

Hess M, Zetouni L, Hess A, Budel J, Dodds K, Henry H, Brauning R, Mcculloch A, Hickey S, Johnson P, Elmes S, Wing J, Bryson B, Knowler K, Hyndman D, Baird H, McRae K, Jonker A, Janssen P, McEwan J, Rowe S. 2022. Combining host and rumen metagenome profiling for selection in sheep: prediction of methane, feed efficiency, production, and health traits. Research Square.

Hess MK, Rowe SJ, Van Stijn TC, Henry HM, Hickey SM, Brauning R, Mcculloch AF, Hess AS, Kirk MR, Kumar S, Pinares-Patiño C, Kittelmann S, Wood GR, Janssen PH, McEwan J. C. 2020. A restriction enzyme reduced representation sequencing approach for low-cost, high-throughput metagenome profiling. PLOS ONE 15: e0219882.

Hu X, Liu G, Shafer ABA, Wei Y, Zhou J, Lin S, Wu H, Zhou M, Hu D, Liu S. 2017. Comparative Analysis of the Gut Microbial Communities in Forest and Alpine Musk Deer Using High-Throughput Sequencing. Frontiers in Microbiology 8.

Keum GB, Pandey S, Kim ES, Doo H, Kwak J, Ryu S, Choi Y, Kang J, Kim S, Kim HB. 2024. Understanding the Diversity and Roles of the Ruminal Microbiome. Journal of Microbiology 62: 217-230.

Park T, Cersosimo LM, Radloff W, Zanton GI, Li W. 2022. The rumen liquid metatranscriptome of post-weaned dairy calves differed by pre-weaning ruminal administration of differentially-enriched, rumen-derived inocula. Animal Microbiome 4: 4.

Parks DH, Chuvochina M, Rinke C, Mussig AJ, Chaumeil PA, Hugenholtz P. 2021. GTDB: an ongoing census of bacterial and archaeal diversity through a phylogenetically consistent, rank normalized and complete genome-based taxonomy. Nucleic Acids Research 50: D785-D794.

Rabee AE, Mohamed M, Ghandour M, Sallam A, Elwakeel EA, Mohammed RS, Sabra EA, Abdel-Wahed AM, Mourad DM, Hamed AA, Hafez OR. 2024. Rumen fermentation and microbiota in Shami goats fed on condensed tannins or herbal mixture. BMC Veterinary Research 20: 35.

Wang X, Chen Y, Shang Y, Wu X, Wei Q, Chen J, Yan J, Zhang H. 2019. Comparison of the gut microbiome in red deer (Cervus elaphus) and fallow deer (Dama dama) by high-throughput sequencing of the V3-V4 region of the 16S rRNA gene. ScienceAsia 45: 515.

New Zealand Journal of Animal Science and Production 2024 Vol 84. 23-25

BRIEF COMMUNICATIONS: Effect of feeding a hyper-diverse pasture mix on pre-weaning ewe and lamb performance

GV Gronqvist[a], LM Cranston[a*], FJ Roca Fraga[a], NW Sneddon & PR Kenyon[a]

[a] School of Agriculture and Environment, Massey University
*Corresponding author: l.cranston@massey.ac.nz

Abstract

This study investigates the effect of feeding a hyper-diverse pasture mix on pre-weaning ewe and lamb performance. Ewes were allocated to either standard pasture (ryegrass/red and white clover) or a hyper-diverse pasture mix (containing 21 species) at set-stocking and grazed that pasture type until weaning. Ewe and lamb live weights, lamb survival and pasture masses were monitored throughout the period. There was no difference (P>0.05) in lamb live weight at birth, docking or lamb survival between those that grazed the standard pasture compared to the hyper-diverse pasture. At weaning, there was no difference (P>0.05) in the live weight of ewes which grazed the two pasture types, however, lambs which grazed the hyper-diverse pasture were heavier than lambs which grazed the standard pasture. There was no difference (P>0.05) in average pasture mass of the two pasture types at a given timepoint (set-stocking, docking, post-docking, weaning). However, at docking the average pasture mass in the hyper-diverse pasture dropped below 1200 kg DM/ha which likely resulted in restricted feed intake of ewes and lambs. This study suggests indicates more research is required to determine the combined effect of pasture masses, botanical composition and pasture quality in hyper-diverse pasture mixes on ewe and lamb performance during the lactation period.

Introduction

Standard perennial ryegrass and white clovers pasture covers for sheep has been well characterized in mid-to late-pregnancy and lactation and suggest that there are no production advantages in offering pasture masses above 1200 kg DM/ha (Morris et al. 1993). Multiple bearing and rearing ewes often, however, fail to meet their nutritional energy demand as seen in a reduced body condition (BCS), this can adversely affect both ewe and lamb performance (Kenyon et al. 2014).

Recently there has been an increased interest in the use of diverse pastures (Cranston et al. 2015; Pembleton et al. 2015). These previous studies have demonstrated that alternative herbages may allow for greater dry matter intake or have greater nutritional qualities resulting in increased ewe milk production and ewe and lamb weaning weight (Grace et al. 2019; Hutton et al. 2011; Kenyon et al. 2012). This study investigated the effect of feeding a hyper-diverse pasture (DP) mix (containing 21 species) between set-stocking and weaning on pre-weaning ewe and lamb performance compared to ryegrass/white clover pasture (SP).

Materials and methods

Study design

This study was carried out at the Massey University, Pasture and Crop Research Unit, in Palmerston North. The pasture types were established in Autumn 2022 after full cultivation and sown with a roller drill. Standard pasture included perennial ryegrass, red clover and white clover. Hyper-diverse pasture included 21 species: perennial ryegrass, meadow fescue, cocksfoot, timothy, red clover, white clover, chicory, plantain, lotus corniculatus, sainfain, sheep burnett, balansa clover, persian clover, arrowleaf clover, subterranean clover, strawberry clover, vetch. Once established the pastures were rotationally grazed with sheep throughout winter and two weeks prior to the start of lambing, pregnant ewes were set-stocked at a rate of 15 ewes per hectare for both pasture types (6 hectares each, 12 hectares total). Set-stocking represented the start of the study period.

Pasture measurements

Pasture mass was measured across all paddocks using a rising plate meter (Ashgrove Pastoral Productions, New Zealand, 50 readings per paddock) with a standard calibration (herbage mass = [158×average meter reading]+200; Hodgson et al. 1999) at set stocking (24 Aug 22), docking (10 Oct 22), post docking (7 Nov 22) and weaning (12 Dec 22).

Animal Measurements

Ewes were weighed unfasted, and body condition scored using a 0-5 scale, including half units ((Jefferies 1961); at set stocking (24 Aug 22), docking (10 Oct 22), post docking (07 Nov 22) and at weaning (12 Dec 22).

All lambs were identified to their dam, within 12 hours of birth, weighed, tagged and had their birth rank and sex recorded. Lambs were then brought into the yards with their dams during docking and weaning and weighed unfasted.

Statistical Analysis

All data were analysed with SAS (SAS version 9.4, North Carolina, USA). Pasture mass was analysed for each measurement date using a mixed model with the fixed effect of pasture type and the random effect of block. Ewe live weight and BCS were analysed for each measurement date using a mixed model with the fixed effect of pasture type and included ewe BCS at the start as a covariate. Lamb live weight was analysed for each measurement date using a mixed model with the fixed effect of pasture type, sex of lamb, rearing rank and pasture type*rearing rank interaction, the random effect of ewe and the covariate of lamb age as the deviation from the median age. Lamb survival was analysed as a binomial trait with a logit transformation using the PROC GLIMMIX procedure, which included the random effect of ewe and the fixed effects pasture type, sex of lamb and their interaction.

Results

Pasture measurements

There were no significant differences (P>0.05) between treatments in pasture mass (kg DM/ha) at set stocking (1918 ± 120 vs. 2047 ± 120 kg/ha for DP and SP respectively), docking (1144 ± 120 vs. 1347 ± 120 kg/ha for DP and SP respectively), post docking (1273 ± 120 vs. 1383 ± 120 kg/ha for DP and SP respectively) or weaning (2310 ± 120 vs 2266 ± 120 kg/ha for DP and SP respectively).

Ewes

Between pasture types, there were no differences (p>0.05) in ewe live weight or BCS at set stocking, docking or weaning (Table 1).

Lambs

Between pasture types, there were no differences (P>0.05) in the live weight of lambs at birth or at docking. At weaning, lambs born to ewes grazing standard pasture were heavier (P<0.05) than lambs born to ewes grazing diverse pasture. There were no differences (P>0.05) in the survival of lambs born to ewes grazing standard pasture or diverse pasture (72% (64-79) for standard pasture vs. 67% (59-74) for diverse pasture).

Table 1. Effect of pasture type (Ryegrass/white clover; Standard pasture and Diverse pasture) on ewe live weight (kg) and body condition score (BCS) at set stocking (24 Aug 22), docking (10 Oct 22), and weaning (12 Dec 22).

Time	*n*	Set Stocking	*n*
Liveweight (kg)			
Standard pasture	91	77.6 ± 1.2	67
Diverse pasture	90	76.2 ± 1.2	70
p value		0.417	
BCS			
Standard pasture	91	2.9 ± 0.1	67
Diverse pasture	90	2.8 ± 0.1	70
p value		0.366	

Means within treatments and columns with different superscripts differ significantly ($p<0.05$).

Table 2. Effect of pasture type (Ryegrass/white clover; Standard pasture and Diverse pasture) on lamb live weight (kg) at birth, docking (10 Oct 22), and weaning (12 Dec 22).

Time	*n*	Birth	*n*	Docking	*n*	Weaning
Live weight (kg)						
Standard pasture	197	5.4±0.1	110	20.0 ± 0.6	111	28.2^{a} ± 0.5
Diverse pasture	191	5.2±0.1	119	18.7 ± 0.6	121	25.4^{b} ± 0.5
p value		0.129		0.104		0.001

Means within treatments and columns with different superscripts differ significantly ($p<0.05$).

Discussion

Previous studies under New Zealand grazing conditions have found that feeding multiple bearing and rearing ewes an herb sward mix can improve both ewe and lamb weaning weights (Hutton et al. 2011; Kenyon et al. 2012). However, in those studies the feed allowances provided allowed for unrestricted ewe intake. In the current study the pasture mass offered fell below 1200 kg DM/ha at docking in the diverse pasture treatment, which likely resulted in restricted feed intake of ewes and lambs in that treatment group. Further, the pasture masses were measured using a plate meter and a calibration equation originally designed for a standard ryegrass/white clover mix (Hodgson et al. 1999), and visual observation suggested the actual pasture masses may have been lower than the measured values. A more accurate calibration equation is likely required if farmers are to successfully use these hyper-diverse pasture mixes in their systems to achieve high rates of sheep performance. Consequently, at weaning lamb live weights were almost 3 kg lighter than those which grazed the hyper-diverse pasture compared to the standard pasture, although there were no differences in ewe weaning live weights. Previous evidence shows that the metabolisable energy content of pasture is a key driver of ewe and lamb weaning weights (Kenyon & Webby 2007). Further, pasture quality samples were not collected in the current study, however, evidence indicates increased pasture diversity commonly results in improved pasture quality relative to standard pasture (Pembleton et al. 2015). In the present study, either an improved pasture quality was not apparent, or the restricted pasture covers meant the ewes could not fully utilise this pasture. Combined, this study suggests more research is required to determine the presentation of hyper-diverse pasture mixes required to improve ewe and lamb performance during the lactation period, in terms of pasture masses, botanical composition and pasture quality.

References

Cranston LM, Kenyon PR, Morris ST, Kemp PD 2015. A review of the use of chicory, plantain, red clover and white clover in a sward mix for increased sheep and beef production. Journal of New Zealand Grasslands 77: 89-94.

Grace C, Lynch M, Sheridan H, Lott S, Fritch R, Boland T 2019. Grazing multispecies swards improves ewe and lamb performance. Animal 13: 1721-1729.

Hodgson J, Matthews P, Matthew C, Lucas R 1999. Pasture measurement. pp. 59-66. In: New Zealand pasture and crop science. Eds. White J, Hodgson J. Oxford University Press, Auckland, New Zealand.

Hutton PG, Kenyon PR, Bedi MK, Kemp PD, Stafford KJ, West DM, Morris ST 2011. A herb and legume sward mix increased ewe milk production and ewe and lamb live weight gain to weaning compared to a ryegrass dominant sward. Animal Feed Science and Technology 164: 1-7.

Jefferies BC 1961. Body condition scoring and its use in management. Tasmanian Journal of Agriculture 32: 19-21.

Kenyon PR, Hickson RE, Hutton PG, Morris ST, Stafford KJ, West DM 2012. Effect of twin-bearing ewe body condition score and late pregnancy nutrition on lamb performance. Animal Production Science 52: 483-490.

Kenyon PR, Maloney SK, Blache D 2014. Review of sheep body condition score in relation to production characteristics. New Zealand Journal of Agricultural Research 57: 38-64.

Kenyon PR, Webby RW 2007. Pastures and supplements in sheep production systems. p. 54. In: Pasture and Supplements for Grazing Animals. Eds. Rattray P, Brookes I, Nicol A. New Zealand Society of Animal Production, Hamilton, New Zealand.

Morris ST, Parker WJ, Blair HT McCutcheon SN 1993. Effect of sward height during late pregnancy on intake and performance of continuously stocked June-and August-lambing ewes. Australian Journal of Agricultural Research 44: 1635-1651.

Nicol AM, Brookes IM 2007. The metabolisable energy requirements of grazing livestock. In: Pasture and supplements for grazing animals. Eds. Rattray PV, Brookes IM, Nicol AM New Zealand Society of Animal Production, Hamilton, New Zealand.

Pembleton KG, Tozer KN, Edwards GR, Jacobs JL, Turner LR 2015. Simple versus diverse pastures: opportunities and challenges in dairy systems. Animal Production Science 55: 893-901.

New Zealand Journal of Animal Science and Production 2024 Vol 84. 26-29

BRIEF COMMUNICATION: Can the effects of intrauterine growth retardation on bone morphology be overcome by first calving? - A preliminary study.

MJ Gibson [1*], CW Rogers [1,2], PJ Back [1,2], KE Dittmer [1], AS Wehrle-Martinez [1], DE Dalley [3] & RR Woods [3]

[1] *School of Veterinary Sciences, Massey University, Private Bag 11-222, Palmerston North 4442, New Zealand,*
[2] *School of Agriculture and Environment, Massey University, Private Bag 11-222, Palmerston North 4442, New Zealand,*
[3] *DairyNZ Ltd., PO Box 85066, Lincoln University, Lincoln 7647, New Zealand*

*Corresponding author email: m.gibson@massey.ac.nz

Abstract

Intrauterine growth retardation, reduced bone density, and an increased risk of fracture were observed in a cohort of calves born to dams fed a diet of baleage and either fodder beet or kale during winter. This study aimed to examine whether the intrauterine effects of the dam's diet on bone development can be overcome by the time the calf reaches its first lactation. Heifer calves from these cohorts were grown, mated, calved, and milked under standard farm practice. At 11 weeks post-calving, heifers were slaughtered, and the humerus was collected and scanned using peripheral quantitative computed tomography (pQCT). There were no differences in pQCT measures of bone size, density, and strength between heifers whose dams were wintered on kale or fodder beet before calving ($p>0.05$). Results indicate subsequent bone growth and development after birth may compensate for the intrauterine effects of late gestation dam diet.

Keywords: Fodder beet; Kale; Protein

Introduction

New Zealand dairy farms operate under a pasture-based system, with crops and supplements used to compensate for feed deficits during winter. In the southern regions of New Zealand, fodder beet (*Beta vulgaris*) is a commonly used crop due to its high metabolisable energy, consistent quality, high yield, and low cost per kilogram produced (Waghorn et al. 2018). As a sole diet, fodder beet is not balanced for dairy cows due to its low crude protein, calcium, and phosphorus content (Jonker et al. 2022).

In cows, the majority of intrauterine growth, particularly musculoskeletal growth occurs in the last trimester of pregnancy (Mao et al. 2008). With 85% of herds in New Zealand being spring-calving, the last trimester of pregnancy most commonly occurs in winter when pasture yield and quality are low (Edwards et al. 2021). If fodder beet is not supplemented correctly to provide adequate protein and minerals, intrauterine growth retardation of bone can occur (Dittmer et al. 2017; Hammond et al. 2022). An earlier study reported that calves born to dams fed (wintered on) fodder beet while pregnant had reduced bone density and strength, and an increased fracture risk compared to calves born to dams fed kale (Gibson et al. 2024).. The aim of this study was to examine whether the intrauterine effects of the dam's diet on bone development can be overcome by the time the calf reaches its first lactation.

Materials and methods

This study was conducted with approval from the Ruakura Animal Ethics Committee, Hamilton, New Zealand (RAEC #14453 and #15106 (lactation)).

This study followed a second cohort of calves from a previous study that examined the intrauterine effects of maternal diet (fodder beet vs. kale, supplemented with baleage) on calf stature and bone at birth(Gibson et al. 2024). Calves from the first cohort were euthanised within 10 days of birth and a forelimb and section of rib including the costochondral junction were harvested for peripheral quantitative computed tomography and histological analysis. There was no difference in calf birth weight, wither height and crown rump length between calves born to each dam diet. Calves born to dams on the fodder beet treatment had reduced bone density and strength compared to calves from the kale treatment. In addition, calves born to dams wintered on FB had smaller girth size than calves born to dams wintered on kale (Gibson et al. 2024).

Calves were born in August-September 2020 at the Southern Dairy Hub research farm (Branxholme, Southland), from dams fed a winter diet of either kale (kale treatment) or fodder beet (fodder beet [FB] treatment), with both groups also having access to pasture baleage. Heifers in the current study (kale = 18 and FB = 17) were reared, mated, calved, and milked under standard farm practice in one cohort. Liveweight was recorded 17 times from birth until approximately 19 months of age. Linear measurements (to the nearest cm) and liveweight (to the nearest kg) were taken within two days of birth. Linear measures of wither height, girth, wither-rump length (wither to tuber ischii) and leg length (from ground to point of olecranon) were obtained.

As calves, these heifers were fed on milk (once daily after the first two weeks), pasture and meal until they reached weaning weight (100 kg), after which they continued to graze pasture and were supplemented with calf meal. At six months of age, heifers were moved to a grazier and reared under standard farm practice. Pasture silage was allocated during times of low pasture availability (i.e. winter).

Heifers returned to the Southern Dairy Hub at the beginning of May 2022 and were transitioned onto a diet consisting of 8.7 kg DM of kale and 3.3 kg DM of baleage. Two weeks prior to their expected calving date, heifers were transported to a commercial farm and were offered a grass and baleage diet with minerals (400 g/cow/day of Transition Cow Advanced [Agvance Nutrition, Howick, NZ] and 15-40 g/cow/day of SoluPhos® [260 g P/kg; Agvance Nutrition, Howick, NZ]). Ten days before the expected calving date, heifers were moved to a calving pad with self-feed silage, hay, and straw, and continued to receive the transition mineral mix and SoluPhos®.

After calving, heifers were provided *ad libitum* grass and baleage was spread with an 80/20 calcium magnesium mix [containing calcium carbonate and magnesium oxide; Healthy Cow, AB LIME, Winton, NZ]. Heifers were milked once-a-day for eight days and offered an in-shed feed consisting of 0.33 kg/cow/day dried distillers grain, 0.33 kg/cow/day crushed grain, 0.33 kg/cow/day palm kernel expeller (PKE), and 165 g/cow/day of OptiPrill (Agvance Nutrition, Howick, NZ). After approximately eight days of once-a-day milking, the heifers were moved into the main milking mob and milked twice a day. They were offered approximately 23 kg DM/cow/day, consisting of grass with 1 kg/cow/day dried distillers grain, 1 kg/cow/day crushed grain, and 1 kg/cow/day PKE, plus 165 g/cow/day OptiPrill.

At approximately 11 weeks post-calving (26 months of age), heifers were slaughtered at a commercial abattoir and the left humerus was collected and sent to Massey University (Palmerston North, Manawatū) for subsequent analysis.

pQCT bone parameters

Peripheral quantitative computed tomography (pQCT; XCT 2000; Stratec Medical, Pforzheim, Germany) scanning was on the humerus (Gibson et al. 2021). Briefly, pQCT scanning was carried out using an XCT 2000 peripheral quantitative computed tomography machine (Stratec Medical, Pforzheim, Germany). For each humerus, a 2 mm slice at the mid-diaphysis was obtained with a voxel size of 0.3 mm. Total bone length was defined as the distance between the proximal end of the humeral head at the lateral aspect, to the end of the trochlea at the distal end. The mid-diaphysis was defined as 50% of the total bone length, starting from the proximal end of the trochlea. Within the manufacturer's software, voxels >710 mg/cm^3 calcium hydroxyapatite (CaHA) were assigned as "cortical" bone. Data derived from the scan included measures of total bone content, periosteal circumference, endosteal circumference, and stress–strain index (SSI). The SSI is a derived measure of bone strength using pQCT data and avoids the requirement for destructive testing using techniques such as the three-point bending test. The SSI describes the ability of bone to withstand bending from lateral, dorso-palmar, and torsional forces and is calculated by incorporating the index of material stiffness (bone mineral density) and bone geometry (cross-sectional moment of inertia) (Ferretti 1995). The pQCT variables will be referred to as bone parameters in the statistical models.

Histology

Histology was undertaken using the same protocol outlined in Wehrle-Martinez et al. (2023). Humeri were longitudinally sectioned with a bandsaw, and then 3-4 mm thick slabs of the proximal humerus were obtained. The bone slabs were placed in 10% neutral buffered formalin solution until processing. During procesing, the bone slabs were demineralised using a hydrochloric acid-based product (Decalcifier hydrochloric acid, Amber Scientific Ltd, Midvale, Australia). Once sufficiently demineralised, four sections of bone from the proximal humeral metaphysis were trimmed and placed in tissue cassettes. The bone sections were then processed routinely, embedded in paraffin wax, and sectioned at 4 µm thickness, followed by staining with haematoxylin and eosin (Wehrle-Martinez et al. 2023).

The histology sections were examined microscopically, and images were taken with an Olympus cellSense imaging software (version 4.2). Physeal thickness was measured on section 1, at three random locations across the slide. Cortical thickness was measured on section 2, 10 mm below the bottom of the physis. Cortical resorption was assessed by counting the number of resorption canals in the cortex of section 4. A two-sample t-test was used to determine if there were any significant differences between the two groups of animals.

Statistical analysis

Statistical analysis was conducted using the Statistical Analysis System software version 9.4 (SAS Institute Inc, Carey, NC, USA). All analyses for pQCT bone parameters and histology measures are presented using means and standard error. Differences in bone measures between dam diets were examined using ANOVA with a significance level set at $p<0.05$. Stature measures were anlaysed using a general linear model with the fixed effect of dam diet and covariate of birth weight. Nine heifers were removed from the dataset due to either failing to get pregnant (FB=4, Kale=1) or the humerus was not collected at the abbatoir (FB=4). This left 26 heifers (FB=9, kale= 17) in the analysis.

Results

There was no significant difference in birth weight and average daily gain between treatments (Table 1, $p>0.10$). There was no effect of treament on calf stature measures (height, girth, length).

There was no significant difference in carcass weight, pQCT measures or histology between treatments (Table 1, $p>0.20$).

Discussion

Both the pQCT and histological analysis require the harvesting of the humerus from the animal of interest, as such this prevents serial sampling of the same individual during different stages of growth. In the subset of calves examined at birth, intrauterine growth retardation in the form of reduced bone mineral density and increased fracture risk was observed in the calves from dams fed fodder beet (Gibson et al. 2024). These differences were not present in the heifers from the same cohorts examined in early lactation in this study

Foetal programming in rats has indicating that offspring from dams fed a low protein diet had altered bone structure and density that persisted into later life (Lanham et al. 2008; Mehta et al. 2002). However, the magnitude of growth retardation from nutritional deficiencies is dependent on both the severity and exposure time during gestation (length of exposure and stage of gestation). The heifers in the current study were from the 2020-born cohort of the calf trial that was reported to have a shorter exposure to the nutritional challenge compared with the cohort from the previous year in the same calf trial. The differences in the calves (at birth) (Gibson et al. 2024), between dam diet treatments were less dramatic in 2020 than in 2019, which had a greater number of days of dam nutritional challenge (7.5 weeks vs 11 weeks respectively). Thus, the duration of nutritional challenge exposure may not have been sufficient to induce sustained impairment of humeral bone growth and development in the developing foetus. However, although the material bone properties were not different between the treatments, the tendency (p=0.06) for the humerus to be longer in heifers from the kale treatment indicating that dam diet may have impeded longitudinal growth at some stage.

The results of the current study should be considered preliminary as there are limitations. Researchers were unable to enter the abattoir due to restricted access, thus preventing the identification of heifer carcasses. Although heifers were grouped into treatments during processing, individual heifer ID could not be linked to each carcass. As a result, humeral bone measurements could not be correlated with the liveweight and carcass weight of individual heifers to adjust for size differences. Given measures of the humerus are highly correlated with liveweight, adjusting for animal size would remove some variation in measurements (Gibson et al. 2019). If this study were to be repeated, it would be essential to ensure the identification of the humerus with heifer ID.

Table 1. *Means ± standard error of stature measures, bone parameters in the mid-diaphysis and histology of the humerus from heifers born to dams fed either kale (n=13) or fodder beet (n=13) at 26 months of age.*

	Fodder beet	**Kale**	**p-value**	
Weight measures			Birth weight	Treatment
n	9	17		
Birth weight (kg)	25.2±1.3	26.9±0.9	-	0.28
Average daily gain to 19 months (kg/day)	0.6 ± 0.01	0.6 ±0.01		0.25
Carcass weight (kg)	134.9±4.6	141.2±3.4	-	0.28
Height (cm)	67.5±1.0	69.2±0.8	<0.01	0.21
Girth (cm)	71.0±1.0	70.9±0.7	<0.01	0.92
Length (cm)	54.5±1.3	56.9±0.9	0.54	0.15
pQCT measures				
n	9	17		
Humerus length (mm)	256.2±2.0	261.2±1.6		0.06
Periosteal circumference (mm)	139.1±1.7	141.3±1.4		0.33
Total bone area (mm^2)	1542.0±38.1	1591.2±31.1		0.33
Endosteal circumference (mm)	93.1±1.8	96.6±1.5		0.16
Cortical bone thickness (mm)	7.3±0.1	7.1±0.1		0.24
Cortical bone density (mg/mm^2)	1288.8±5.3	1286.2±4.3		0.36
Cortical bone content (mg/mm)	1093.9±24.2	1088.4±19.7		0.86
Stress-strain index (mm^3)	11774.6±366.2	11988.5±299.0		0.65
Histology				
n	9	17		
Thickness of physis (µm)	470.5±25.5	452.1±18.5		0.56
Cortical thickness (µm)*	828.3±107.0	839.4±70.8		0.93
Number of cortical osteoclastic resorption cavities	9.6±1.7	8.6±1.2		0.67

**Missing 3 measurements due cut bone section not being long enough to measure cortical thickness (FB=2, Kale =1)*

Conclusion

The lack of differences in bone measures observed at two years of age between heifers whose dams were fed either kale or fodder beet prior to calving indicates that a linear growth pattern is sufficient to overcome intrauterine growth limitations. However, this conclusion may only be applicable to the duration and intensity of the dam's nutritional challenges experienced in this trial. It is important to note that the calves were managed and slaughtered under commercial conditions, which imposed logistical constraints that should be considered when interpreting the magnitude of the effect observed in the results. Additional research is required to determine whether the lack of differences in bone morphology at two years of age truly reflects the ability of bone to overcome the intrauterine effects of dam diet.

Acknowledgements

The authors would like to acknowledge Mathew Perrott, Petru Daniels and Evelyn Lupton for histology section preparation, Bryce Ilton and Kirsty Anderson for assistance with limbs transport. We would also like to thank the DairyNZ research technicians and the farm team at the Southern Dairy Hub for animal management and sample collection. A special thank you also to the commercial dairy farm who calved the heifers and took them through their first 11 weeks of lactation. This project was funded by the Sustainable Food and Fibre Futures fund administered by the Ministry for Primary Industries, Wellington, New Zealand (Contract 405512); DairyNZ (Contract CB1923) and PGG Wrightson Seeds.

References

Dittmer K, Morley R, Smith R 2017. Skeletal deformities associated with nutritional congenital rickets in newborn lambs. New Zealand Veterinary Journal 65: 51-55.

Edwards J, Cuthbert S, Pinxterhuis J, McDermott A 2021. The fate of calves born on New Zealand dairy farms and dairy farmer attitudes towards producing dairy-beef calves. New Zealand Journal of Animal Science and Production 81: 179-185.

Ferretti J 1995. Perspectives of pQCT technology associated to biomechanical studies in skeletal research employing rat models. Bone 17: S353-S364.

Gibson M, Dittmer K, Hickson R, Back P, Wehrle-Martinez A, Rogers C 2021. The mid-diaphysis is a poor predictor of humeral fracture risk indicating that predisposing factors are recent. Ruminants 1: 23-30.

Gibson MJ, Rogers CW, Back PJ, Dittmer KE, Wehrle-Martinez A, Dalley DE, Woods RR 2024. The intrauterine effects of a maternal winter diet of either kale or fodder beet on measures of calf stature and bone morphology at birth. New Zealand Journal of Agricultural Research: 1-16.

Gibson MJ, Rogers CW, Dittmer KE, Hickson RE, Pettigrew EJ, Back PJ 2019. Can bone measures of the bovine metacarpus predict humeral bone structure? New Zealand Journal of Animal Science and Production 79: 8-12.

Hammond KJ, Sandoval E, McKenzie CM, Lees S, Pacheco D, McCoard SA 2022. The effect of a fodder beet versus rye-grass grazing regime during mid-to-late gestation twin-bearing ewes on dam and progeny performance and lamb survival. New Zealand Journal of Agricultural Research 65: 145-162.

Jonker A, Dalley D, Woods R, Dynes R, Pacheco D, Khan M, Edwards J, Hammond K 2022. Modelling winter feeding practices with fodder beet in growing and dry dairy cattle. New Zealand Journal of Animal Science and Production 82: 39-44.

Lanham S, Roberts C, Perry M, Cooper C, Oreffo R 2008. Intrauterine programming of bone. Part 2: alteration of skeletal structure. Osteoporosis International 19: 157-167.

Mao W, Albrecht E, Teuscher F, Yang Q, Zhao R, Wegner J 2008. Growth-and breed-related changes of fetal development in cattle. Asian-Australasian Journal of Animal Sciences 21: 640-647.

Mehta G, Roach H, Langley-Evans S, Taylor P, Reading I, Oreffo R, Aihie-Sayer A, Clarke N, Cooper C 2002. Intrauterine exposure to a maternal low protein diet reduces adult bone mass and alters growth plate morphology in rats. Calcified Tissue International 71: 493-498.

Waghorn G, Law N, Bryant M, Pacheco D, Dalley D 2018. Digestion and nitrogen excretion by Holstein–Friesian cows in late lactation offered ryegrass-based pasture supplemented with fodder beet. Animal Production Science 59: 1261-1270.

Wehrle-Martinez A, Lawrence K, Back P, Rogers C, Gibson M, Dittmer K 2023. Osteoporosis is the cause of spontaneous humeral fracture in dairy cows from New Zealand. Veterinary Pathology 60: 88-100.

AUTHOR INDEX

New Zealand Society of Animal Production (Inc.) (NZSAP)
PO Box 955
Cambridge, New Zealand
www.nzsap.org.nz

ISBN 979-8-3313-2213-7

9 798331 322137

CURRAN ASSOCIATES INC.
proceedings.com

International Conference on Information Systems and Management Science (ISMS 2018)

Misida, Malta
22 – 23 February 2018